SpringerBriefs in Applied Sciences and Technology

SpringerBriefs in Computational Intelligence

Series Editor

Janusz Kacprzyk, Systems Research Institute, Polish Academy of Sciences, Warsaw, Poland

SpringerBriefs in Computational Intelligence are a series of slim high-quality publications encompassing the entire spectrum of Computational Intelligence. Featuring compact volumes of 50 to 125 pages (approximately 20,000-45,000 words), Briefs are shorter than a conventional book but longer than a journal article. Thus Briefs serve as timely, concise tools for students, researchers, and professionals.

More information about this subseries at http://www.springer.com/series/10618

Oscar Castillo · Luis Rodriguez

A New Meta-heuristic Optimization Algorithm Based on the String Theory Paradigm from Physics

Oscar Castillo
Division of Graduate Studies and Research
Tijuana Institute of Technology
Tijuana, Mexico

Luis Rodriguez
Division of Graduate Studies and Research
Tijuana Institute of Technology
Tijuana, Mexico

ISSN 2191-530X ISSN 2191-5318 (electronic)
SpringerBriefs in Applied Sciences and Technology
ISSN 2625-3704 ISSN 2625-3712 (electronic)
SpringerBriefs in Computational Intelligence
ISBN 978-3-030-82287-3 ISBN 978-3-030-82288-0 (eBook)
https://doi.org/10.1007/978-3-030-82288-0

This Springer imprint is published by the registered company Springer Nature Switzerland AG
The registered company address is: Gewerbestrasse 11, 6330 Cham, Switzerland

Preface

This book focuses on the fields of bio-inspired algorithm, optimization problems and fuzzy logic. In this book, we proposed a new metaheuristic based on String Theory. It is important to mention that we have proposed new algorithm to generate new potential solutions in optimization problems in order to find new ways that could improve the results in solving these problems; we present the results for the proposed method in different cases of study; the first case is of 13 traditional benchmark mathematical functions, and the second case is the optimization of benchmark functions of the CEC 2015 Competition; with the basic randomness proposed method, we are presenting a set that we selected the first 10 problems of the CEC 2017 Competition on Constrained Real-Parameter Optimizer that are problems with constraints that contain the presence of constraints that alter the shape of the search space making it more difficult to solve. Finally, we are presenting the optimization of a fuzzy inference system (FIS), specifically finding the optimal design of a fuzzy controller. It is important to mention that in all case study, we are presenting statistical tests in order to prove the performance of proposed method.

This book is intended to be a reference for scientists and engineers interested in applying a different metaheuristic in order to solve optimization and applied fuzzy logic techniques for solving problems in intelligent control. This book can also be used as a reference for graduate courses like the following: soft computing, bio-inspired algorithms, intelligent control, fuzzy control and similar ones. We consider that this book can also be used to get novel ideas for new lines of research, new techniques of optimization, or to continue the lines of the research proposed by the authors of the book.

In Chap. 1, we begin by offering a brief introduction of the proposed method, the general features of the metaheuristics and its application; also, we are presenting the inspiration, justification and main contribution of this work.

We describe in Chap. 2 the literature review, basic theoretical and technical concepts about the areas of computational intelligence and the features and characteristics of String Theory that we need to design the String Theory Algorithm. In addition, we presented a brief introduction to the history of the first and main algorithms bio-inspired and their classification.

We describe in more detail in Chap. 3 the proposed method, the String Theory Algorithm (STA), we presented equations, examples, explanation and figures of each step that we developed in order to create a new algorithm. In this section, we can note the inspiration about physical theory and how it is used in the proposed metaheuristic. It is important to mention that in the randomness method, we are presented two different ways to implement and are called basic and advanced randomness proposed method, respectively.

Chapter 4 is devoted to present the results for the proposed method in different case of study; the first case is of 13 traditional benchmark mathematical functions, and the second case is the optimization of benchmark functions of the CEC 2015 Competition; with the basic randomness proposed method, we are presenting a set that we selected the first 10 problems of the CEC 2017 Competition on Constrained Real-Parameter Optimizer that are problems with constraints that contain the presence of constraints that alter the shape of the search space making it more difficult to solve. Finally, we are presenting the optimization of a fuzzy inference system (FIS), specifically finding the optimal design of a fuzzy controller. It is important to mention that in all cases study, we are presenting statistical tests in order to prove the performance of proposed method.

We offer in Chap. 5 the conclusions of this book work are presented at the end in order to mention the advantages of the proposed method and a brief summary of the String Theory Algorithm (STA).

We end this preface of the book by giving thanks to all the people who have helped or encouraged us during the writing of this book. First of all, we would like to thank Esmeralda Rosas and Maximo Rodriguez for their love and always supporting our work and especially for motivating us to write our research work. We would also like to thank our colleagues working in soft computing, which are too many to mention each by their name. Of course, we need to thank our supporting agencies, CONACYT and Tijuana Institute of Technology, in our country for their help during this project. We have to thank our institution, Tijuana Institute of Technology, for always supporting our projects. Finally, we thank our respective families for their continuous support during the time that we spend in this project.

Tijuana, Mexico   Prof. Oscar Castillo
                                                                                    Dr. Luis Rodriguez

Contents

Chapter 1
Introduction

Nowadays it is more common for researchers in the area of computer science to work with bio-inspired metaheuristics to solve optimization problems [1]. There are many of papers and algorithms that researchers have developed based on natural behaviors that have been recently proposed. The main challenge in developing a new algorithm is to represent and to simulate as much as possible each of the most important behaviors and characteristics of real phenomena. This representation can be implemented by operators in the corresponding algorithms, for example, in genetic algorithms [2] we use operators, like crossover, mutation and selection.

In this work, we are presenting a new algorithm inspired on the theoretical basis and behavior of String Theory in physics [3]. In this case, as part of this proposal we also are presenting a new approach to randomness based on the behavior of the vibrations that are produced by each string in the universe. In other words, a new method to generate random numbers inside the new algorithm is proposed. In addition, we are presenting the motion equations, which the algorithm needs to find the best result. Finally, we are presenting a newly proposed method for performing crossover on strings in the algorithm in order to improve the solutions, and this proposed method is based on real numbers in order to represent the natural combination occurring in Strings in nature.

It is important to mention that in this work we have proposed new algorithm to generate new potential solutions in optimization problems in order to find new ways that could improve the results in solving these problems. In addition, we are presenting new ways to generate randomness that we use in other algorithms and prove that only the proposed randomness method works and improve the performance of these algorithms. Finally we presented a new crossover operator respectively that could be implemented in other metaheuristics in order to potentially improve their performance. The main difference of this new algorithm in comparison with others that exist in the literature is that this proposed method takes the best features

© The Author(s), under exclusive license to Springer Nature Switzerland AG 2022
O. Castillo and L. Rodriguez, *A New Meta-heuristic Optimization Algorithm Based on the String Theory Paradigm from Physics*,
SpringerBriefs in Computational Intelligence,
https://doi.org/10.1007/978-3-030-82288-0_1

and advantages of String Theory in developing a computer science model and new techniques that other algorithms do not use, but can be modified in order to include the new techniques that we have proposed in this research.

The main contribution of this work is the new techniques that we are presenting in order to generate new STA approach for solving optimization problems. These techniques are mainly the following: a new randomness method in order to implement the main feature of the stochastic methods that are the randomness factor method, the second proposed method is the combination among Strings, and this combination is used with the real numbers and is based on the box plot. The third proposed technique is the combination (in the algorithm) of two laws that are very similar but with an important variation which is the mathematical sign (positive and negative), these techniques are the electromagnetics and the gravity laws respectively. These techniques are based on String Theory. Finally, with this inspiration, we can conclude that the simulation of this theory in computer science can improve the performance of set problems that we analyzed in this work.

The book is structured as follows:

In Chap. 2, we describe the literature review, basic theoretical and technical concepts about the areas of computational intelligence and the features and characteristics of String Theory that we need to design the String Theory Algorithm. In addition, we presented a brief introduction to the history of the first and main algorithms bio-inspired and their classification.

In Chap. 3, we describe in detail the proposed method, String Theory Algorithm (STA) we presented equations, examples, explanation and figures of each step that we develop in order to create a new algorithm. In this section, we can note the inspiration about physical theory and how it is used in the proposed metaheuristic. It is important to mention that in the randomness method we are presented two different ways to implement and are called basic and advanced randomness proposed method respectively.

In Chap. 4, we present the results for the proposed method in different case of study, the first case is of 13 traditional benchmark mathematical functions, the second case is the optimization of benchmark functions of the CEC 2015 Competition, with the basic randomness proposed method we are presenting a set that we selected the first 10 problems of the CEC 2017 Competition on Constrained Real-Parameter Optimizer that are problems with constraints that contain the presence of constraints that alter the shape of the search space making it more difficult to solve. Finally, we are presenting the optimization of a fuzzy inference system (FIS), specifically finding the optimal design of a fuzzy controller. It is important to mention that in all cases study we are presenting statistical tests in order to prove the performance of proposed method.

In Chap. 5, the conclusions of this book work are presented at the end in order to mention the advantages of the proposed method and a brief summary of the String Theory Algorithm (STA).

References

1. B. Melián, J. Moreno, Metaheurísticas: una visión global. Revista Iberoamericana de Inteligencia Artificial **19**, 7–28 (2003)
2. A. Thengade, R. Donal, Genetic algorithm—survey paper, in *IJCA Proceedings, National Conference on Recent Trends in Computing, NCRTC*, vol. 5 (2012)
3. J. Polchinski, *String Theory*, vol. 1 (Cambridge University Press, Cambridge, 2001)

Chapter 2
Literature Review

In this section, we describe the literature review that was used in this book in order to have basic concepts and information about computational intelligence, bioinspired algorithms and different techniques that the researchers use in optimization problems.

2.1 Metaheuristics

Optimization is everywhere, be it engineering design or industrial design, business planning or holiday planning, etc. We use optimization techniques to solve problems intelligently by choosing the best from a larger number of available options. Metaheuristics have earned more popularity over exact methods in solving optimization problems because of the simplicity and robustness of results produced while implemented in widely varied fields including engineering, business, transportation, and even social sciences. There is established extensive research by metaheuristic community, which involves the introduction of new methods, applications, and performance analysis. However, Srensen [1] believes that the field of metaheuristics has still to reach maturity as compared to physics, chemistry, or mathematics. There is immense room of research to appear on various issues faced by metaheuristic computing.

2.2 Classification of Metaheuristics

Nowadays, researchers study and propose new techniques in order to find the particular algorithm or meta-heuristic that solves the greatest number of problems,

© The Author(s), under exclusive license to Springer Nature Switzerland AG 2022
O. Castillo and L. Rodriguez, *A New Meta-heuristic Optimization Algorithm Based on the String Theory Paradigm from Physics*, SpringerBriefs in Computational Intelligence, https://doi.org/10.1007/978-3-030-82288-0_2

although, the NFL (No Free Lunch) Theorem [2] has logically proven that there is no meta-heuristic appropriately suited for solving all optimization problems. For example, a particular meta-heuristic can provide very promising results for a set of problems, but the same algorithm can show poor performance in another set of different problems. Therefore we can find a diversity of techniques and new meta-heuristics that are being created for solving optimization problems. Finally, we can mention that meta-heuristic techniques can be classified as follows:

- Evolutionary (based on the evolution concept in nature) [3]: Genetic Algorithms (GA) [4, 5], Evolutionary Programming (EP) [6] and Genetic Programming (GP).
- Based on Physics (imitate the rules of physics) [7]: Big-Bang Big Crunch (BBBC) [8], Gravitational Search Algorithm (GSA) [9] and Artificial Chemical Reactions Optimization Algorithm (ACROA) [10].
- Swarm Intelligence (Social Behavior of swarms, herds, flocks or schools of creatures in nature) [11, 12]: Particle Swarm Optimization (PSO) [13], Ant Colony Optimization (ACO) [14] and the Bat-inspired Algorithm (BA) [15].

2.3 Genetic Algorithm

A Genetic Algorithm (GA) is possibly the first algorithm that was developed in order to simulate genetic systems. This algorithm was first proposed by Fraser [16, 17] and later by Bremermann [18] and Reed in [19]. Finally, Holland [20] did extensive work on Genetic Algorithms that popularized GAs, and this is a main reason that Holland is generally considered as the father of GAs. Basically, genetic algorithms were inspired on genetic evolution, where the features of individuals are expressed using genotypes. The main operators used in this algorithm are selection, crossover and mutation, respectively and we can find more details in [5, 21].

2.4 Particle Swarm Optimization

Another popular algorithm is the particle swarm optimization (PSO) algorithm, which is a population-based search algorithm based on the simulation of the social behavior of birds within a flock. The initial goal of the particle swarm concept was to graphically simulate the graceful and unpredictable choreography of a bird flock [13], this with the aim of discovering patterns that govern the ability of birds to fly synchronously and to suddenly change direction with regrouping in an optimal formation. From this initial objective, the concept evolved into a simple and efficient optimization algorithm.

The main equations of the Particle Swarm Optimization that Kennedy and Eberhart [13] originally proposed that the position x^i per each particle i, is evolved as:

$$x_{k+1}^i = x_k^i + v_{k+1}^i \qquad (2.1)$$

and its velocity v^i is calculated as:

$$v_{k+1}^i = v_k^i + c_1 \times r_1 \times \left(p_k^i - x_k^i\right) + c_2 \times r_2 \times \left(p_k^g - x_k^i\right) \qquad (2.2)$$

The subscript k, indicates the increment of a time. p_k^i represents the best position of the particle i at time k so far, while p_k^y is the global best position in all particles in the cognitive and social scaling parameters respectively which are selected such that $C_1 = C_2 = 2$ to give a mean equal 1 when they are multiplied by r_1 and r_2. The using of these values make the particles overshooting the target in the half of time. Equation 2.2 is used to determinate the new velocity v_{k+1}^i of the i-th particle's, at time k, while Eq. 2.1 gives the i-th particle's new position x_{k+1}^i by adding its v_{k+1}^i to its current position x_k^i.

2.5 Grey Wolf Optimization

The Grey Wolf Optimizer (GWO) algorithm [22] is a metaheuristic of quite recent creation, in 2014 Mirjalili Seyadali created this algorithm inspired in the hunting behavior of the grey wolf. Basically, the phenomena that GWO imitates are the hunting mechanics and the social hierarchy of the grey wolf. The GWO has two main operations that we can describe below:

$$\boldsymbol{D} = C\boldsymbol{X}_p(t) - \boldsymbol{X}(t) \qquad (2.3)$$

where $\boldsymbol{D}$ is the Manhattan distance that exists between a randomness movement C of the best wolf $\boldsymbol{X}_p(t)$ and the current wolf $\boldsymbol{X}(t)$ respectively and this equation simulates the hunting mechanics. It is important to say that in this algorithm the randomness C is a random number between [0, 2]. In addition the next position of the current wolf is represented by the following equation:

$$x(t+1) = \boldsymbol{X}_p(t) - A\boldsymbol{D} \qquad (2.4)$$

where $x(t+1)$ shows the next position of current wolf, $\boldsymbol{X}_p(t)$ is the best wolf in the population, A is the weight distance (scalar number) that was calculated above $\boldsymbol{D}$. In other words the movement of each wolf is the subtraction between the best solution and the distance between the current wolf and the best wolf with randomness and applying a weight in this distance. Finally Mirjalili represented the social hierarchy with the following equation:

$$x(t+1) = \frac{X_1 + X_2 + X_3}{3} \qquad (2.5)$$

where X_1, X_2 and X_3 are the same values of Eq. 2.4, but in this case they are based on the three best wolves that are called alpha, beta and delta respectively [23]. In the original paper the hierarchical pyramid is used as a centroid method and as a different way to implement other existing methods [24].

2.6 Firefly Algorithm

Firefly Algorithm (FA) [25] created in 2008 by Xin-She Yang and was inspired by the flashing patterns and behavior of fireflies. This algorithm has basically three rules that we can find in more detail in the original paper. In this algorithm we can find the two main equations, the first is the attractiveness or the light intensity of each firefly and is represented by the following equation:

$$\beta = \beta_0 e^{-\gamma r^2} \qquad (2.6)$$

where β_0 and γ are parameters that could be optimized according to the problem and r represents the distance among two fireflies, and with this equation the attractiveness of each firefly is presented. Finally the movement of a firefly is calculated with the following equation:

$$x_i^{t+1} = x_i^t + \beta_0 e^{-\gamma r_{ij}^2} \left(x_j^t - x_i^t\right) + \alpha_t \epsilon_i^t \qquad (2.7)$$

where x_i^{t+1} is the new position of the firefly, x_i^t is the current position of the firefly, where we are adding the attractiveness that exists between the other best solution x_j^t and the current solution x_i^t and finally adding the randomness parameter that is α_t being the randomization parameter, and ϵ_i^t is a vector of random numbers drawn from a Gaussian distribution or uniform distribution at time t. In other words, the new position is the current position plus the attractiveness between the current solution and other best solution, and finally adding the randomness value that is a uniform distribution multiplied by the alpha parameter.

2.7 Flower Pollination Algorithm

Flower Pollination Algorithm (FPA) [26] was created by Xin-She Yang inspired by the pollination process of flowers. In the original paper we can find in detail the inspiration and the justification of the process for each operation in the algorithm. Here we

are only presenting the main equations and a brief explanation about the algorithm. Basically, FA presents two different ways to apply the pollination among the flowers (solutions), the first method is called a global pollination and it is represented by the following equation:

$$x_i^{t+1} = x_i^t + L\left(x_i^t - g_*\right) \tag{2.8}$$

where x_i^t represents the vector of i solutions at iteration t, L is a Levy distribution and finally g_* is the best solution in the population, in other words the new position of one flower is the current position plus the distance between the current position and the best solution, multiplied by a randomness factor that in this case is represented by Levy flights.

The second method of movement of the algorithm is called local pollination and is simulated by the following equation:

$$x_i^{t+1} = x_i^t + \epsilon\left(x_j^t - x_k^t\right) \tag{2.9}$$

where x_j^t and x_k^t are pollen from two different flowers that we are evaluating and ϵ is a uniform distribution from 0 to 1, in other words the new position of the current individual is the current position plus the distance among two flowers, which are not necessarily the best ones into the population, multiplied by a uniform distribution.

2.8 Fuzzy Logic

In this book we are presenting an optimization of a fuzzy controller in order to test the performance of the new proposed algorithm (STA), but, is important to have a formal definition of fuzzy logic and the formal definition of fuzzy sets was proposed by Zadeh in 1965 [27] and it is as follows: Let X be a space of points (objects), with a generic element of X denoted by x. Thus, $X = \{x\}$. A fuzzy set A in X is characterized by a membership (characteristic) function, $f_A(x)$ which associates with each point in X a real number in the interval [0, 1], with the value of $f_A(x)$ at x representing the "degree of membership" of x in A. Thus, the nearer the value of $f_A(x)$ to unity, the higher the degree of membership of x in A. When A is a set in the ordinary sense of the term, its membership function can take on only two values 0 and 1, with $f_A(x) = 1$ or 0 according as x does or does not belong to A respectively. Thus, in this case $f_A(x)$ reduces to the familiar characteristic function of a in set A. (When there is a need to differentiate between such sets and fuzzy sets, the sets with two-valued characteristic functions will be referred to as ordinary sets or simply sets.)

References

1. K. Srensen, M. Sevaux, F. Glover, A history of metaheuristics, in *ORBEL29–29th Belgian conference on operations research* (2017)
2. D.H. Wolpert, W.G. Macready, No free lunch theorems for optimization. IEEE Trans. Evolut. Comput. 67–82 (1997)
3. H. Maier, Z. Kapelan, Evolutionary algorithms and other metaheuritics in water resources: current status, research challenges and future directions. Environ. Model. Softw. **62**, 271–299 (2014)
4. F. Aladwan, M. Alshraideh, M. Rasol, A genetic algorithm approach for breaking of simplified data encryption standard. Int. J. Secur. Its Appl. **9**(9), 295–304 (2015)
5. Lingaraj and Haldurai, A study on genetic algorithms and its applications. Int. J. Comput. Sci. Eng. **4**, 139–143 (2016)
6. X. Yao, Y. Liu, G. Lin, Evolutionary programming made faster. IEEE Trans. Evolut. Comput. **3**, 82–102 (1999)
7. U. Can, B. Alatas, Physics based metaheuristic algorithms for global optimization. Am. J. Inform. Sci. Comput. Eng. **1**, 94–106 (2015)
8. K. Osman, E. Ibrahim, A new optimization method: big bang-big crunch. Adv. Eng. Softw. **37**, 106–111 (2006)
9. E. Reshedi, H. Nezamabadi-Pour, S. Saryazdi, GSA: a gravitational search algorithm. Inf. Sci. **179**, 2232–2248 (2009)
10. B. Alatas, ACROA: artificial chemical reaction optimization algorithm for global optimization. Exp. Syst. Appl. **38**, 13170–13180 (2011)
11. G. Beni, J. Wang, *Swarm Intelligence in Cellular Robotic Systems, Robots and Biological Systems: Towards a New Bionics?* (Springer, Berlin, 1993), pp. 703–712
12. X.-S. Yang, M. Karamanoglu, Swarm intelligence and bio-inspired computation: an overview, in *Swarm Intelligence and Bio-inspired Computation* (2013), pp. 3–23
13. J. Kennedy, R.C. Eberhart, Particle swarm optimization, in *Proceedings of the IEEE International Joint Conference on Neural Networks* (1995), pp. 1942–1948
14. M. Dorigo, M. Birattari, T. Stutzle, Ant colony optimization. IEEE. Comput. Intell. Magaz. 28–39 (2006)
15. X.S. Yang, *A New Metaheuristic Bat-Inspired Algorithm* (2010)
16. A. Frases, Simulation of genetic systems by automatic digital computers I: introduction. Austr. J. Biol. Sci. **10**, 484–491 (1957)
17. A. Frases, Simulation of genetic systems by automatic digital computers II: effects of linkage on rates of advance under selection. Austr. J. Biol. Sci. **10**, 492–499 (1957)
18. H.J. Bremermann, Optimization through evolution and recombination, in *Self-organization Systems*, ed. by M.C. Yovits, G.T. Jacobi, G.D. Goldstine (1962), pp 93–106
19. J. Reed, R. Toombs, N.A. Barricelli, Simulation of biological evolution and machine learning. J. Theor. Biol. **17**, 319–342 (1967)
20. J.H. Holland, *Adaptation in Natural and Artificial Systems* (University of Michigan Press, Ann Arbor, 1975)
21. A. Thengade, R. Donal, Genetic algorithm—survey paper, in *IJCA Proceedings, National Conference on Recent Trends in Computing, NCRTC*, vol. 5 (2012)
22. S. Mirjalili, M. Mirjalili, A. Lewis, Grey Wolf Optimizer. Adv. Eng. Softw. **69**, 46–61 (2014)
23. C. Muro, R. Escobedo, L. Spector, R. Coppinger, Wolf-pack (Canis lupus) hunting strategies emerge from simple rules. Comput. Simul. Behav. Process. **88**, 192–197 (2011)
24. L. Rodriguez, O. Castillo, J. Soria, P. Melin, F. Valdez, C. Gonzalez, G. Martinez, J. Soto, A fuzzy hierarchical operator in the grey wolf optimizer algorithm. Appl. Soft Comput. **57**, 315–328 (2017)
25. X.-S. Yang, *Firefly Algorithm: Recent Advances and Applications* (2013), arXiv:1308.3898v1
26. X-S. Yang, *Flower Pollination Algorithm for Global Optimization* (2012), arXiv:1312.5673v1
27. L. Zadeh, Fuzzy sets. Inf. Control **8**, 338–353 (1965)

Chapter 3
String Theory Algorithm

This chapter, a detailed description of the algorithm proposed method is presented.

The main goal of this book is to create a new metaheuristic based on the String Theory and their features, in this chapter we are presenting the proposed method and their equations that were used in order to solve optimization problems, in following sections we describe in detail each step and procedure of String Theory Algorithm.

In addition, we can mention that in classical physics the elementary particle is the atom and with this particle all matter that exists in the universe can be formed and explained. However, String Theory proposes that the universe is formed by the energy vibrations produced by particles called strings. Presently we understand that physics can be described by four forces: gravity, electromagnetism, weak forces, responsible for beta decays and strong forces, which bind quarks into protons and neutrons [1]. Finally, String Theory aims at unifying these four laws and explains the physical phenomena in very large bodies and also at the quantum level. In Table 3.1 we can find the main features of these forces, and the relation and behavior in the String Theory Algorithm.

3.1 Initialization

In the optimization problems is very common that exists a search range, sometimes this range could be the same for every dimension that the problem has or the search range is different for each dimension of the optimization problem. We can define the optimization problem as the following equation:

$$f(x_i) \in R, \; x_i\_min \le x_i \le x_i\_max \tag{3.1}$$

© The Author(s), under exclusive license to Springer Nature Switzerland AG 2022
O. Castillo and L. Rodriguez, *A New Meta-heuristic Optimization Algorithm Based on the String Theory Paradigm from Physics*,
SpringerBriefs in Computational Intelligence,
https://doi.org/10.1007/978-3-030-82288-0_3

Table 3.1 Features of string theory

Force type	Description	Mediating particle	Behavior in the algorithm
Electromagnetism	It includes the laws of electricity and magnetism	Photon	Exploration
Strong nuclear force	It holds the quarks together with the protons and the neutrons and these together in turn to form the atomic nucleus	Gluon	Exploitation
Weak nuclear force	Only force that are able to change the nature of the particle	W and Z bosons	Combination of strings
Gravity	Attract massive bodies (Quantum level behavior)	Graviton	Exploitation

In other words, we can mention that the optimization problems have a lower and upper bound or search range. In the STA, the initialization is in randomness way with the restriction that each string (solution) is in the search range.

3.2   Randomness Method

In this book we are presenting two ways to generate the randomness numbers in the algorithm, this method is very important in the metaheuristics. We called the first method as Basic and is the randomness proposed method based on a simple sine and cosine equation. The second proposed method is called Advanced and this proposed method is based on a Furrier series with some adaptations that we describe in detail in the following sections.

3.3   Basic Method

The first randomness method as we mention above is the basic and is the sine waves method that basically is inspired in a behavior of general sine or cosine waves that we can find in nature. For example, we can find this behavior in the ocean waves that have this important natural behavior or in the behavior in the sound propagation when we are speaking or when we are listening to music. These waves have the behavior of a set of cycles of a sine function or out of phase as the cosine function.

This behavior simulates the propagation of waves and it is represented with the following equation:

$$y(x.t) = A \, \sin\left(\frac{2\pi}{L}x + \frac{2\pi}{T}t\right) \tag{3.2}$$

where the function is defined in space and time, $y(x, t)$ is a function representing position x at time t, A represents the amplitude of the wave (including positive and negative numbers), parameter L is the length of wave, in other words we can say that these are the number of dimensions that the problem has, and finally T is the period that the wave has. In addition the period can be described by the following equation:

$$T = \frac{L}{v} \tag{3.3}$$

where v is the velocity with which the wave is propagated and L is the length of the wave, which was described above. It is important to say that in this case the time (t) represents the number iterations, generations, epochs or other type of time used in each metaheuristic and the space (x) is the position from 1 to the number of dimensions in incremental order.

Figure 3.1 shows an example of a wave generated with amplitude 2 and 30 dimensions, in this case for each dimension we are assigning a value that is represented by the wave (blue line) and in this case the vector of random values is presented in Table 3.2.

In Fig. 3.2 we represent in a graphical way the values that correspond to each dimension according to Eq. 3.2.

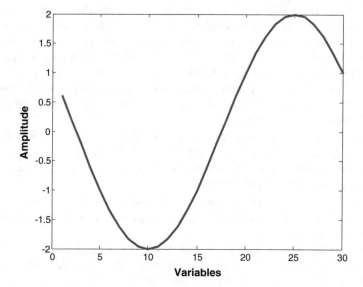

Fig. 3.1 Graphical way of sine wave

Table 3.2 Vector values of sine wave

0.6180	0.2091	−0.2091	...	1.6180	1.3383	1.0000

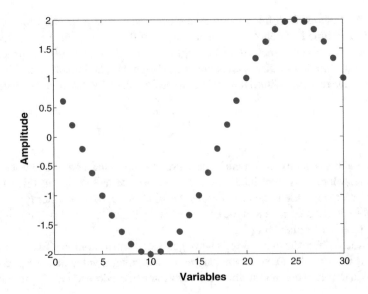

Fig. 3.2 Values in a new method

In addition it is important to mention that the positive and negative numbers help to change the direction of the randomness movement and create greater diversity in order to help find an optimal solution and to avoid local optima. We can clearly notice this behavior when the curve changes between decreasing to increasing.

According to Eq. 3.2 the velocity can be described as a variation of values in the wave at time t, iterations, generations, epochs or other as mentioned above. In order to explain in more detail the importance of the velocity parameter, we are presenting in Figs. 3.3, 3.4 and 3.5 the wave at velocity 1, 5 and 10 at time 1, 2 and 3 respectively with a problem of 30 dimensions.

Figure 3.3 shows the wave with velocity 1, where the blue dots are the values for time 1, the green dots are the values in time 2 and finally the yellow dots represent the values of the wave in time 3, and we can notice the movement of the wave with velocity 1 in three different times.

In Fig. 3.4 we can note the variation of the values when the velocity of the wave is 5, in this case we can note that the values change in a fast way in comparison to when the case of velocity is 1.

Finally, in Fig. 3.5 we can find the values in times 1, 2 and 3 when the velocity is 10, and in this case the values are changing faster between times 1 and 2. In other words the velocity is a parameter that affects directly in the variety or diversity of the values in the sine wave between the current time and the next time, so according to the problem that we need to solve, this parameter (velocity) is important for the performance of the algorithm.

In addition it is important to mention that the proposed method is recommended when the problem has more than 10 variables or dimensions, and the main reason

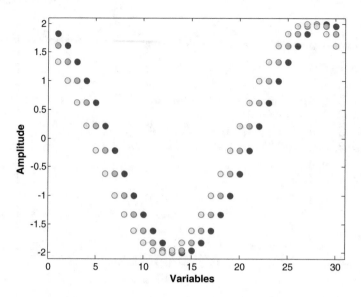

Fig. 3.3   Sine wave with velocity 1

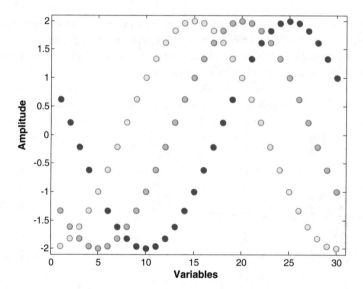

Fig. 3.4   Sine wave with velocity 5

is because the proposed method needs points as reference to correctly draw the sine wave. In the following figures we can appreciate in a graphical way the reason for this suggestion.

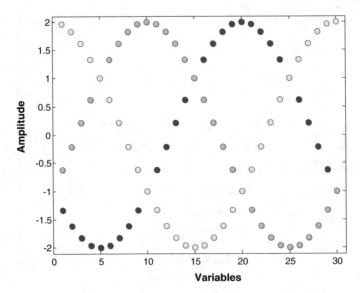

Fig. 3.5 Sine wave with velocity 10

Figure 3.6 shows the points and numbers that the proposed method generates when the problem has only 3 dimensions and we can note that the sine wave lost the features that we described above and the form that we can draw is very similar to a triangle when we use 3 variables or dimensions.

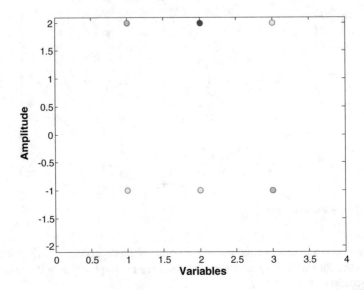

Fig. 3.6 Points for 3 variables

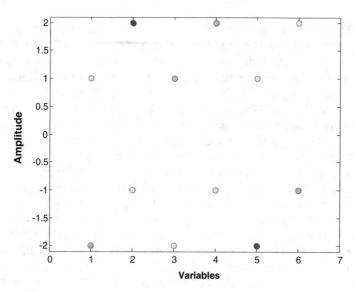

Fig. 3.7 Points for 6 variables

Also we present in the graphical way in Fig. 3.7 the points that we can draw when the problem has only 6 dimensions or variables.

According to the figures presented in this book, we can conclude that to use the proposed method is important to analyze if the problem that we need to solve, has more than 10 dimensions in order to satisfy the features that are necessary to implement and draw a correct sine wave, and this is an important suggestion so that the proposed method works and has a good performance.

Finally, we are presenting in this book a brief literature review of similar researchs that uses the sine cosine laws.

In 2016 Mirjalili S. published the Sine Cosine Algorithm (SCA) [2] for solving optimization problems, in this specific work the author used a mathematical model based on sine and cosine functions in order to move the potential solutions into the search space and the SCA is used to find solutions in problems as selected best features/attributes to improve the classification accuracy [3, 4]. Also we can find the SCA for solving real parameters bound constrained single objective optimization problems [5] or in multi-objective engineering design problems [6]. The SCA has also been used in hybridizations such as Sine Cosine Crow Search Algorithm [4] or GWO-SCA [7] in order to create a powerful combination to solve complex optimization problems. It is important to mention that the main difference between the SCA optimizer and the proposed method is that the first one is an optimization algorithm and the proposed method in this work is a new method to generate stochasticity or randomness into the algorithm, so the proposed method can be adapted in other metaheuristics for improving their performance. Finally the details of this work are presented in the Proposed Method section.

In addition other authors have studied and implemented changes to the methods to generate random numbers (stochastic method) in the algorithms that we mentioned above. Yang [8] changed the Gauss distribution to Levy Flights as a method to generate random numbers into the Firefly Algorithm and concluded that the modification helps considerably in the performance of the algorithm. Heiradi [9] also uses Levy Flights in the GWO algorithm in order to improve the performance of the algorithm, the comparison was applied in a set of classic benchmark functions and concluded that the Algorithm improves significantly in comparison to the original method to generate randomness. Yang and Heidari changed the method for another one that already exists in the literature and showed a good performance, and this is an inspiration to the proposed new method to generate stochasticity or randomness that can be adapted in the metaheuristics to help improve their performance.

In recent works researchers that apply metaheuristics to solve optimization problems have used algorithms like the Genetic Algorithm (GA), Particle Swarm Optimization (PSO) [10], Ant Colony Optimization (ACO), Artificial Bee Colony (ABC) [11] or Fireworks Algorithm (FWA) [12, 13] as the most popular among the metaheuristics currently available in computer science.

3.4   Advanced Method

In the second randomness method we use as much as possible the String Theory. In this case, we are assuming that all strings are open strings, and we did not include closed strings, for simplicity. In this way, we can represent and approximation the Strings vibration as a Fourier expansion in space and time as the following equation:

$$y(x, t) = \sum_{n=1}^{\infty} B_\beta \cos \omega_n t \, \sin \frac{n\pi x}{L} \tag{3.4}$$

where n is a normal model of the oscillating system, B_β represents the amplitude of successive harmonics, L is the length of the string and finally w_n is represented by the following equation:

$$\omega_n = v \frac{n\pi}{L} \tag{3.5}$$

where v is representing the velocity generated in the string according to the tension, mass and other features of the String.

In Fig. 3.8 we can find an example of Eq. 3.4 and we can note an interesting behavior and is a clear over shoot just at the beginning of the Fourier series. In order to generate random numbers with the same probability we proposed to remove this over shoot by using only a specific range of values within the series, as we can note in Fig. 3.9.

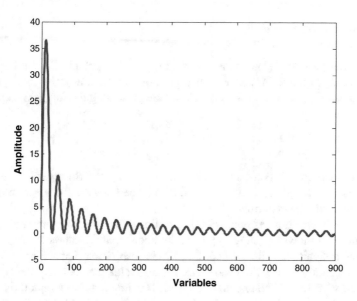

Fig. 3.8 Wave equation based on Fourier series

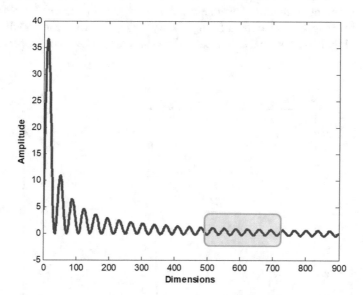

Fig. 3.9 Specific range of Eq. 3.4

It is important to mention that in the algorithm these values are normalized and are used with amplitude in order to easily adjust to the features that the particular problem requires. This is the proposed method that we are presenting in order to generate the randomness in the algorithm based on String Theory.

3.5   String Movement

On the other hand as we mentioned above, the gravitational and Coulomb laws are very similar, the main difference is that one can produce a negative result and the other one is only positive. So in this case we are presenting the following equation:

$$F = K\frac{q_1 q_2}{r^2} \tag{3.6}$$

where K is Coulomb's constant, q_1 is the fitness of the best String, q_2 is the fitness of the current string and finally, r_2 is the distance between the position of each dimension of these two strings.

Figure 3.10 shows an example of this equation with a random sample.

Figure 3.10 shows an interesting behavior and is that we can note that only one value has a significant force (according to the inspiration that we are using). In other words, only in one dimension we have one value that can help the algorithm to achieve diversity in the exploration phase. It is important to mention that in the algorithm these values are ranked and are used with amplitude in order to easily adjust with the features that the problem requires. This is the proposed method that we are presenting in order to generate diversity in the exploration and exploitation phases of the algorithm based on String Theory. In Fig. 3.11 we can find the same values of Fig. 3.10, but now in this case, the values are ranked with respect to the amplitude.

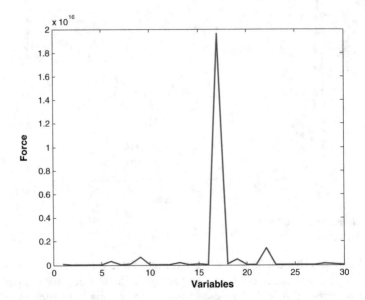

Fig. 3.10 Coulomb's law

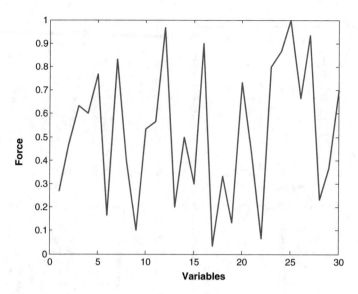

Fig. 3.11 Coulomb's law ranked

Finally, In Fig. 3.12 we can find more examples of this behavior.

In order to represent the features mentioned above we have proposed the following equation that we are using to move the positions of the strings in the proposed algorithm.

$$x(t+1) = x(t) + \text{sgn}(t)\left(\left(|x_b(t) - x(t)|K\frac{f(t)f_b(t)}{|x_b(t) - x(t)|^2}\right)\right.$$
$$\left. + \left(\sum_{n=1}^{\infty} B_\beta \cos(\omega_n t) \sin\left(\frac{n\pi x}{L}\right)\right)\alpha\right) \tag{3.7}$$

where $x(t)$ represents the string in the current iteration, $x_b(t)$ is the best solution in the algorithm, the Coulomb's law is present in the multiplication with the distance and explained in more detail in Eq. 3.6 and finally, the last part of the equation represents the randomness method that we have proposed in Eq. 3.4 (advanced proposed randomness method).

$$\text{sgn}(t) = \begin{cases} -1; \; if \; p < 0.5 \\ 1; \; if \; p \geq 0.5 \end{cases} \tag{3.8}$$

Equation 3.8 is for switching between the electromagnetic and gravitational forces, respectively (changing the sign), where p is a probability that is a random number between 0 and 1.

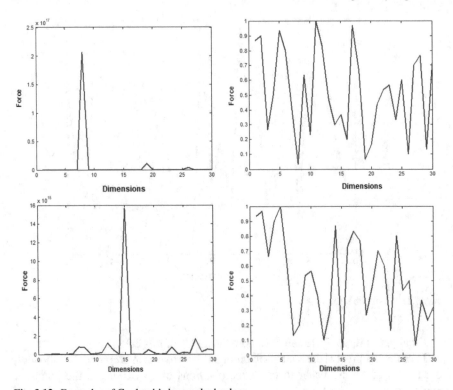

Fig. 3.12 Examples of Coulomb's law ranked values

3.6 Boxplot Crossover

In addition, we are presenting a proposed method in order to achieve combinations among different Strings (solutions in the algorithm). These combinations are representing new possible solutions when the algorithm has potential solutions in a range of the continuous numbers.

In this case, we have proposed generating two new Strings based on the combination between the best solution and another random string in the population. So with this method, we have the two parents that we need to generate the proposed combined method (Combination of Strings based on the box plot method) in order to improve the population in a stochastic way.

In order to show an example of this method, we are presenting the random values of one position (dimension) of the two Strings in the Algorithm, and it is important to remember that the proposed method is represented with continuous values.

Table 3.3 shows the values for two Strings in one specific dimension in order to have a combination to create two new Strings in the algorithm.

The next step in this method is generating at least 31 random numbers in the range of the parents [1.25–99.85]. In addition, it is important to order these numbers in

Table 3.3 Values in strings combination

Parents	Values
Parent 1	99.85
Parent 2	1.52

Table 3.4 The 35 Random numbers in the range of the parents

16.63	17.82	18.00	18.05	18.20	19.86	21.15	27.36	29.32	29.99	32.80	33.52	40.31
41.79	42.34	45.76	46.55	47.36	47.36	50.58	52.99	53.90	54.39	56.77	58.97	60.85
62.47	71.56	73.77	75.06	77.11	80.10	80.17	83.40	84.24				

an increasing fashion in order to generate a box plot. In Table 3.4 we can find this example with 35 random numbers.

The elements of the box plot are: minimum value, percentile 25, percentile 50, percentile 75 and the maximum value. In order to know the percentiles 25 and 75, respectively in the random numbers, we are using the following equation:

$$L = \frac{k}{100}(n) \tag{3.9}$$

where L is the position that results from the equation and corresponds to the value into the vector of random numbers, k is the percentile that we are searching for and n is the total of random numbers that we generated above.

In this case, we need the percentiles 25 and 50, respectively. Then by using Eq. 3.9 we have the following numbers as the results: the positions 9 and 27 respectively. This means that we can take the values in these positions of Table 3.4, as the new results of the combination among two strings.

In Table 3.5 we can find the first new value in italic and the second one in bold.

In Figs. 3.13 and 3.14 respectively, we can find the surface and box plot of this proposed method.

In Table 3.6 we can find the results that the proposed Combination String method produced, and we can conclude that now we have two strings based on the parents and we can note that these two new strings are similar to the originally selected parents.

Table 3.5 The new values based on plot box

16.63	17.82	18.00	18.05	18.20	19.86	21.15	27.36	*29.32*	29.99	32.80	33.52	40.31
41.79	42.34	45.76	46.55	47.36	47.36	50.58	52.99	53.90	54.39	56.77	58.97	60.85
62.47	71.56	73.77	75.06	77.11	80.10	80.17	83.40	84.24				

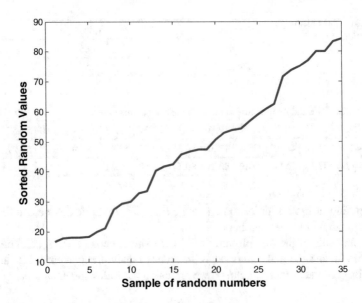

Fig. 3.13 Sample of sorted random values based on two strings

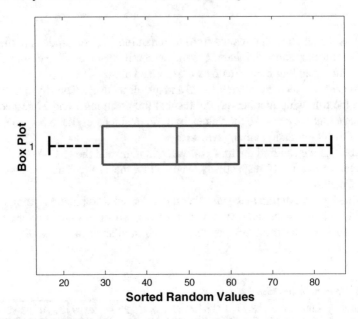

Fig. 3.14 Box plot of proposed combination strings method

Table 3.6 Results of the proposed method

Offspring	Values
Offspring 1	62.47
Offspring 2	29.32

3.7 Pseudocode and Flowchart

In Fig. 3.15 we can find the pseudocode of the String Theory Algorithm based on the main equations that we described above.

Figure 3.16 shows the Flowchart of the String Theory Algorithm (STA).

Initialize the String Population X_i $(i = 1, 2, \ldots, n)$
Initialize B_β, p and α
Calculate the fitness of each Strings
X_b = the best search agent
while (t < maximum number of iterations)
    for each search agent
        Apply the Box plot combination operator
        if (p > 0.5)
            Update the position of the current search agent
            by equation (3.7) using (+) adding operator
        else
            Update the position of the current search agent
            by equation (3.7) using (-) subtraction operator
        end if
    end for each
    calculate the fitness of each Strings
    Update X_b
    t = t + 1
end while
Return X_b

Fig. 3.15 Pseudocode of string theory algorithm

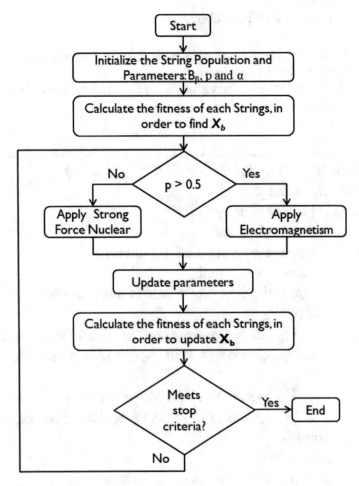

Fig. 3.16 Flowchart of string theory algorithm

References

1. J. Polchinski, *String Theory*, vol. 1 (Cambridge University Press, Cambridge, 2001)
2. S. Mirjalili, SCA: A Sine Cosine algorithm for solving optimization problems. Knowl.-Based Syst. **96**, 120–133 (2016)
3. A.I. Hafez, H.M. Zawbaa, E. Emary, A.E. Hassanien, Sine cosine optimization algorithm for feature selection. International Symposium on Innovations in Intelligent Systems and Applications (2016), pp 1–5
4. N. Singh, S.B. Singh, A novel hybrid GWO-SCA approach for optimization problems. Eng. Sci. Technol. Int. J. (2017)
5. R. Kommadath, J. Dondeti, P. Kotecha, Benchmarking JAYA and Sine Cosine algorithm in real parameter bound constrained single objective optimization problems (CEC 2016). International Conference on Intelligent Systems, Metaheuristics & Swarm Intelligence (2017), pp. 31–34
6. M.A. Tawhid, V. Savsani, Multi-objective Sine-Cosine algorithm (MO-SCA) for multi-objective engineering design problems. Neural Comput. Appl. 1–15 (2017)

7. M. Dorigo, M. Birattari, T. Stutzle, Ant colony optimization. IEEE. Comput. Intell. Mag. 28–39 (2006)
8. X.-S. Yang, M. Karamanoglu, Swarm intelligence and bio-inspired computation: an overview, in *Swarm Intelligence and Bio-Inspired Computation* (2013), pp. 3–23
9. A. Heiradi, P. Parham, An efficient modified grey Wolf optimizer with Levy flight for optimization tasks. Appl. Soft. Comput. **60**, 115–134 (2017)
10. J. Soto, P. Melin, O. Castillo, Particle swarm optimization of the fuzzy integrators for time series prediction using ensemble of IT2FNN architectures, in *Nature-Inspired Design of Hybrid Intelligent Systems*, eds. By P. Melin, O. Castillo, J. Kacprzyk. Studies in Computational Intelligence, vol. 667 (Springer, Berlin, 2017), pp. 141–158
11. B. Basturk, D. Karaboga, An artificial bee colony (ABC) algorithm for numeric function optimization, in IEEE, *Swarm Intelligence Symposium* (2006), pp. 12–14
12. Y. Tan, *Fireworks Algorithm* (Springer, Berlin, 2015), pp. 355–364
13. J. Barraza, P. Melin, F. Valdez, C. Gonzalez, Fuzzy fireworks algorithm based on a spark dispersion measure. Algorithms **10**(3), (2017)

Chapter 4
Simulation Results

In this chapter, we present results obtained of the String Theory Algorithm with the different study cases that we are presented in the problem definition section. In addition, we show the statistical tests of the study cases in order to prove the performance of the String Theory Algorithm with the hypothesis test.

4.1 Definition Problems

In order to test the String Theory Algorithm, we are presenting the following problems that we used in this book.

4.1.1 First Case Study: Traditional Benchmark Functions

For this book, we will specifically evaluate the proposed method for 13 benchmark functions[1–3]. These benchmark functions are classified as unimodal and multimodal and are functions that have been already tested in other metaheuristics, especially to test new algorithms and verify that they have a good performance.

The functions are described below in Tables 4.1 and 4.2 respectively, where "Range" represents the boundary of the function's search space and f_{min} is the optimal value. We also present plots of the 3-D versions of the benchmark functions in Figs. 4.1 and 4.2 with the goal of analyzing the form and search space of the functions that we are going to test with the proposed method.

The main reason to test the proposed approach with these benchmark functions is because there exists a great variety of metaheuristics that use these benchmark

© The Author(s), under exclusive license to Springer Nature Switzerland AG 2022
O. Castillo and L. Rodriguez, *A New Meta-heuristic Optimization Algorithm Based on the String Theory Paradigm from Physics*,
SpringerBriefs in Computational Intelligence,
https://doi.org/10.1007/978-3-030-82288-0_4

Table 4.1 Unimodal benchmark functions

Name function	Mathematical definition	Range	f_{min}
Sphere	$f_1(x) = \sum_{i=1}^{n} x_i^2$	$[-100, 100]$	0
Schwefel 2.22	$f_2(x) = \sum_{i=1}^{n}|x_1| + \prod_{i=1}^{n}|x_1|$	$[-10, 10]$	0
Rotated hyper-ellipsoid	$f_3(x) = \sum_{i=1}^{n}\left(\sum_{j-1}^{i} x_j\right)^2$	$[-100, 100]$	0
Schwefel 2.21	$f_4(x) = \max_i\{|x_1|, 1 \le i \le n\}$	$[-100, 100]$	0
Rosenbrock	$f_5(x) = \sum_{i=1}^{n-1}\left[100(x_{i+1} - x_i^2)^2 + (x_1 - 1)^2\right]$	$[-30, 30]$	0
Modified Zakharov	$f_6(x) = \sum_{i=1}^{n}([x_1 + 0.5])^2$	$[-100, 100]$	0
Quartic	$f_7(x) = \sum_{i=1}^{n} i x_i^4 + random[0, 1]$	$[-1.28, 1.28]$	0

functions to validate their performance and mainly because it is important to test the proposed method in this work under the same conditions that other metaheuristics.

Table 4.1 and Fig. 4.1 show the equations with their characteristics and the 3-D versions of the unimodal benchmark functions respectively, and these benchmark functions are evaluated with different dimensions according to the analyzed metaheuristic.

In Table 4.2 and Fig. 4.2 we can find the equations with their characteristics and the 3-D versions of the multimodal benchmark functions respectively.

4.1.2   Second Case Study: Benchmarks Functions of the CEC 2015 Competition

In addition the proposed method was also tested with more complex benchmark functions, specifically with benchmark functions that were presented in the CEC 2015 competition [4]. In Table 4.3 we can find brief specifications of these functions and the main features are that the search range is now from -100 to 100, and the number of function evaluations to solve the problem are limited in order to ensure fair competition and the maximum numbers are the following: for 10 dimensions = 100,000, for 30 dimensions = 300,000, for 50 dimensions = 500,000 and for 100 dimensions = 1,000,000.

Finally it is important to mention that these are the numbers of dimensions (10, 30, 50 and 100) that we can use testing in these types of problems with and the results are shown as the error regarding the optimal value, in other words, the final results have to be closer to zero.

Table 4.2 Multimodal benchmark functions

Name function	Mathematical definition	Range	f_{min}
Schwefel	$f_8(x) = \sum_{i=1}^{n} -x_i \sin(\sqrt{|x_i|})$	$[-500, 500]$	$-2,094.91$
Rastrigin	$f_9(x) = \sum_{i=1}^{n} \left[x_i^2 - 10\cos(2\pi x_i) + 10 \right]$	$[-5.12, 5.12]$	0
Ackley	$f_{10}(x) = 20 \exp\left(-0.2 \sqrt{\frac{1}{n} \sum_{i=1}^{n} x_t^2} \right)$ $- \exp\left(\frac{1}{n} \sum_{i=1}^{n} \cos(2\pi x_i) \right) + 20 + e$	$[-32, 32]$	0
Griewank	$f_{11}(x) = \frac{1}{4000} \sum_{i=1}^{n} x_1^2 - \prod_{i=1}^{n} \cos\left(\frac{x_i}{\sqrt{i}} \right) + 1$	$[-600, 600]$	0
Levy	$f_{12}(x) = \frac{\pi}{n} \{ 10 \sin(\pi y_1)$ $+ \sum_{i=1}^{n-1} (y_i - 1)^2 \left[1 + 10 \sin^2(\pi y_{y+1}) \right]$ $+ (y_n - 1)^2 \} + \sum_{i=1}^{n} u(x_p, 10, 100, 4)$ $y_1 = 1 + \frac{x_i + 1}{4}$ $u(x_1, a, k, m) = \begin{cases} k(x_i - a)^m, & x_i > a \\ 0, & -a < x_i < a \\ k(-x_i - a)^m, & x_i < -a \end{cases}$	$[-50, 50]$	0
Levy No. 13	$f_{13}(x) = 0.1 \{ \sin^2(3\pi x_i)$ $+ \sum_{i=1}^{n} (x_1 - 1)^2 \left[1 + \sin^2(3\pi x_1 + 1) \right]$ $+ (x_n - 1)^2 \left[1 + \sin^2(2\pi x_n) \right] \}$ $+ \sum_{i=1}^{n} u(x_t, 5, 100, 4)$	$[-50, 50]$	0

4.1.3   Third Case Study: Constrained Functions of the CEC 17 Competition

In this book, we are also presenting problems with constraints that contain the presence of constraints that alter the shape of the search space making it more difficult to solve. In order to test the proposed method in problems with constraints, we selected the first 10 problems of the CEC 2017 Competition on Constrained Real-Parameter

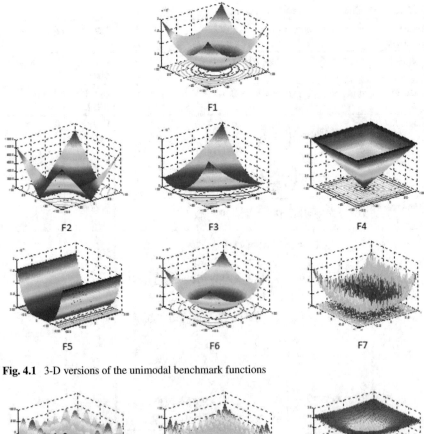

Fig. 4.1 3-D versions of the unimodal benchmark functions

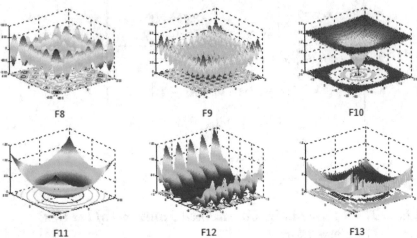

Fig. 4.2 3-D versions of the multimodal benchmark functions

Table 4.3 Summary of the CEC 15 learning-based benchmark suite

	No	Functions	Obj
Unimodal functions	1	Rotated high conditioned elliptic function	100
	2	Rotated cigar function	200
Simple multimodal functions	3	Shifted and rotated Ackley's function	300
	4	Shifted and rotated Rastrigin's function	400
	5	Shifted and rotated Schwefel's function	500
Hybrid functions	6	Hybrid function 1 (N = 3)	600
	7	Hybrid function 2 (N = 4)	700
	8	Hybrid function 3(N = 5)	800
Composition functions	9	Composition function 1 (N = 3)	900
	10	Composition function 2 (N = 3)	1000
	11	Composition function 3 (N = 5)	1100
	12	Composition function 4 (N = 5)	1200
	13	Composition function 5 (N = 5)	1300
	14	Composition function 6 (N = 7)	1400
	15	Composition function 7 (N = 10)	1500

Search range: $[-100, 100]^{\wedge}D$

Optimizer [5]. Table 4.4 shows a brief description of these 10 functions, we can note that the functions have the search range going from -100 to 100, and the number of the function evaluations to solve the problem are limited in order to ensure fair competition and the maximum numbers are the following: Maximum function evaluations $= 20{,}000 * D$; where D is the dimensionality of the optimization problems and the population is free into the algorithm while not exceeding the Maximum number of function evaluations.

4.1.4 Fourth Case Study: Fuzzy Control of an Autonomous Mobile Robot

Finally, we are presenting the last case study that consists on designing a fuzzy controller [6] for an autonomous mobile robot system. The main goal of this problem is to provide the control actions to the motors to minimize the error with respect to a pre-established trajectory. In this case, the optimal fuzzy controller was obtained by the proposed String Theory Algorithm.

In Fig. 4.3 we can find the configuration of the robot, which has 2 wheels with motors and a passive wheel for the stabilization of the system.

The operation of the autonomous mobile robot model is represented by the following equations [7–9]:

Table 4.4 Details of 10 test problems

Problem	Type of objective	Number of constraints	
		E	I
C1	Non separable	0	1; Separable
C2	Non separable, rotated	0	1; Non separable, rotated
C3	Non separable	1; Separable	1; Separable
C4	Separable	0	2; Separable
C5	Non separable	0	2; Non separable, rotated
C6	Separable	6; Separable	0
C7	Separable	2; Separable	0
C8	Separable	2; Non separable	0
C9	Separable	2; Non separable	0
C10	Separable	2; Non separable	0

Search range: $[-100, 100]^D$

D is the number of decision variables, I is the number of inequality constraints, E is the number of equality constraints

Fig. 4.3 Autonomous mobile robot model

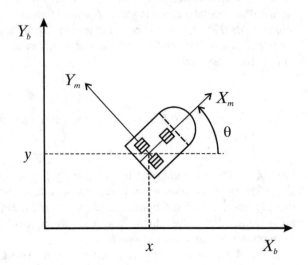

$$M(q)\dot{v} + C(q, \dot{q})v + Dv = \tau + P(\tau) \tag{4.1}$$

$$\dot{q} = \begin{bmatrix} \cos\theta & 0 \\ \sin\theta & 1 \\ 0 & 1 \end{bmatrix} \begin{bmatrix} v \\ w \end{bmatrix} \tag{4.2}$$

where $q = (x, y, \theta)^T$ is the coordinate vector, which describes the position of the robot, $v = (v, w)^T$ is the linear and angular velocity vector, $\tau = (\tau_1, \tau_2)$ is the torque vector applied to the wheels of the robot, where τ_1 and τ_2 represent the right and left wheels, $P \in R^2$ is the uniform disturbance vector, $M(q) \in R^{2x2}$ is a symmetric and positive matrix, $C(q, \dot{q})v$ is the vector of centripetal and Coriolis forces, and $D \in R^{2x2}$ is a diagonal positive defined damping matrix [10].

The fuzzy system of the mobile robot controller consists of two input variables and two output variables respectively. The first input variable is called e_v (error in linear velocity) and the second input variable is called e_w (error in angular velocity). Both input variables have the same membership functions that are two trapezoidal membership functions at both ends and a triangular at the center, labeled as N (negative), Z (zero) and P (positive) [11, 12], respectively.

On the other hand, the two outputs also are similar; the two variables are called τ_1 (torque 1) and τ_2 (torque 2) respectively and have three triangular membership functions labeled N (negative), Z (zero) and P (positive) respectively. In Fig. 4.4 we can find the design of the fuzzy controller for the autonomous mobile robot[13–15].

In Fig. 4.5 we can note the rules that are used to control the dynamics of the autonomous mobile robot problem.

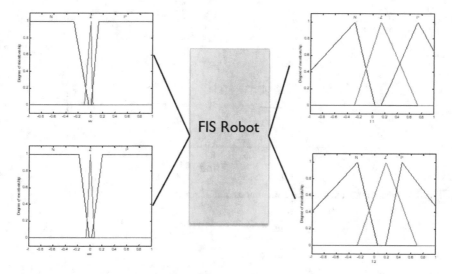

Fig. 4.4 Fuzzy system for the fuzzy controller

1) If (ev is N) and (ew is N) then (TI is N) (T2 is N)
2) If (ev is N) and (ew is Z) then (TI is N) (T2 is Z)
3) If (ev is N) and (ew is P) then (TI is N) (T2 is P)
4) If (ev is Z) and (ew is N) then (TI is Z) (T2 is N)
5) If (ev is Z) and (ew is Z) then (TI is Z) (T2 is Z)
6) If (ev is Z) and (ew is P) then (TI is Z) (T2 is P)
7) If (ev is P) and (ew is N) then (TI is P) (T2 is N)
8) If (ev is P) and (ew is Z) then (TI is P) (T2 is Z)
9) If (ev is P) and (ew is P) then (TI is P) (T2 is P)

Fig. 4.5 Fuzzy rules for the autonomous mobile robot

Finally, we are presenting the desired trajectory for the robot to follow, which begins at the point (0,0) and turns to generate a "U" shaped trajectory, in order to create a more complex problem the trajectory is repeated in several occasions as shown in Fig. 4.6.

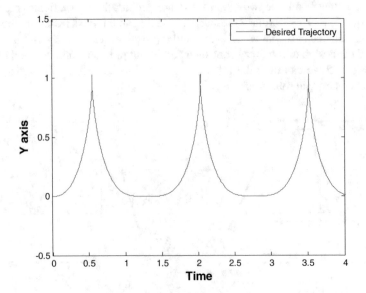

Fig. 4.6 Desired trajectory for the autonomous mobile robot

4.2 Definition of the Comparison Method and Type of Statistical Test

For this book, we are also presenting a statistical test in order to mathematically compare the new proposed algorithm with other metaheuristics. In the tables presented below the hypothesis tests comparing each optimization method with the proposed method are presented. In this case, each table shows the averages and standard deviations obtained in 51 independent runs of each method and finally the obtained values of z in hypothesis testing are presented [16].

The goal is to statistically demonstrate that the new method (STA) obtains better results than the other presented algorithms. We consider the Z-test because in order to realize this test we need a minimum 31 samples and we have 51 samples. So we performed this test with a 95% of confidence level and 5% of significance ($\alpha = 0.05$) and according to the tables of critical values of the Z-test with $\alpha = 0.05$, the range for rejecting the null hypothesis is in $Z_0 = (0, -1.645]$.

The parameters for the hypothesis testing are as follows:

- $\mu_1 = mean\ of\ String\ Theory\ Algorithm$
- $\mu_2 = mean\ of\ the\ algorithm\ proposed\ in\ order\ to\ compare$
- The mean of the new method (STA) is less than the mean of the proposed method (claim).
- $H_0 : \mu_1 \geq \mu_2$
- $H_a : \mu_1 < \mu_2 (Claim)$
- $\alpha = 0.05$
- sample size (n) = 51
- $Z_0 = -1.645$

In addition we need the value of Z to realize the comparison with Z_0 and conclude if there is enough evidence to state that the new method is better than an existing one.

The value of Z is obtained by the following equations: The value of Z is obtained by the following equations:

$$\sigma_{\bar{x}_1 - \bar{x}_2} \approx \sqrt{\frac{s_1^2}{n_1} + \frac{s_2^2}{n_2}} \tag{4.3}$$

$$Z \approx \frac{(\bar{x}_1 - \bar{x}_2)}{\sigma_{\bar{x}_1 - \bar{x}_2}} \tag{4.4}$$

where:

s_1, s_2: are the standard deviations of the two methods respectively.
n_1, n_2: are the sizes of the sample of the methods.
$\bar{x}_1, \bar{x}_2$: are the averages obtained for each method respectively.

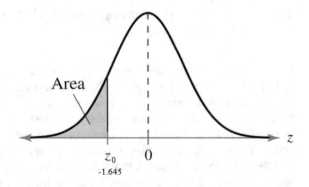

Figure 4.7 shows the rejection region of the null hypothesis (H_0), so if the value of Z is less than Z_0 or -1.645, we can then reject the null hypothesis and we can say that the mean of the new method (STA) is lower than mean of the other methods that were used to compare this work with a 95% of confidence. The values of Z are shown in the following tables to conclude in this way if the new String Theory Algorithm (STA) is better and has a better performance than the other methods. Finally, is important to mention that in all tables that we are presenting in the Simulation Results section have bold results in each row, these bold results represent the algorithm or method that has better performance according to the hypothesis test that we presented in this section.

4.3 Basic Randomness Proposed Method

In this section, we are presenting the adaptation of three algorithms (GWO, FA and FPA) with the basic randomness proposed method, in order to verify the performance of the proposed method. In addition, we present the results and hypothesis test of each algorithm with some study cases.

4.3.1 Adaptation of Grey Wolf Optimizer

In the GWO algorithm [17] the randomness has a particular method that is a simple randomness value that we can find in Eq. 4.5.

$$D = C X_p(t) - X(t) \qquad (4.5)$$

where C is a randomness value that is in the range [0, 2] and represents the stochastic method in the GWO algorithm, and in this case we are including the proposed method in the following Equation:

$$D = SX_p(t) - X(t) \tag{4.6}$$

where S is the proposed method with amplitude 2, in this specific algorithm we can add the absolute method in order to respect the original values or to keep the original sine wave with the negative numbers and to help the algorithm not only to generate randomness values also generate a change of direction of each wolf for this case.

4.3.1.1 Comparison with the First Case Study

For the first case study is important mention that for the GWO algorithm, the parameters for the conventional benchmark functions are the same that the original author of the suggested method (C in the range [0, 2] and "a" parameter in decrement from 2 to 0 through the iterations), we used 30 wolves and 500 iterations for the first set of benchmark functions.

It is important to mention that this algorithm uses the simple randomness as a random number in a range from 0 to 2. In Table 4.5 we are presenting the hypothesis test between the original method denoted as GWO and the proposed method based on the sine waves denoted as GWO-SW and we tested both methods in a first set of benchmark functions with 30 dimensions respectively and we can note that in this case the proposed method is better than the conventional method only in 2 of the 13 benchmark functions that were analyzed.

In order to test the methods with a higher level of complexity, we increase the number of dimensions in the first 13 benchmark functions and in this case in Table 4.6 we can find the results of the hypothesis test when the problems have 64 dimensions

Table 4.5 Comparison between the original and proposed method in the GWO with 30 Dimensions

30 Dimensions					
Function	GWO	STD	GWO-SW	STD	Z-Value
F1	**6.59E−28**	**6.34E−05**	5.90E−45	2.12E−44	−5.69E−23
F2	**7.18E−17**	**0.0290**	9.54E−27	1.36E−26	−1.36E−14
F3	**3.29E−06**	**79.1496**	1.72E−07	7.62E−07	−2.16E−07
F4	**5.61E−07**	**1.3151**	1.32E−12	3.56E−12	−2.34E−06
F5	26.8126	69.9050	27.4607	0.6406	0.0508
F6	**0.8166**	**1.26E−04**	1.6351	0.4269	10.5018
F7	**0.0022**	**0.1003**	1.00E−03	8.69E−04	−0.0663
F8	−6123.10	4087.44	**−4395.29**	**1298.42**	−2.2066
F9	0.3105	47.3561	0.00	0.00	−0.0359
F10	**1.06E−13**	**0.0778**	1.01E−14	2.89E−15	−6.75E−12
F11	0.0045	0.0067	**0.00**	**0.00**	−3.6890
F12	**0.0534**	**0.0207**	0.0909	0.0329	5.2763
F13	**0.6545**	**0.0045**	0.9688	0.2375	7.2479

Table 4.6 Comparison between the original and proposed method in the GWO with 64 Dimensions

64 Dimensions					
Function	GWO	STD	GWO-SW	STD	Z-value
F1	8.56E−17	7.90E−17	**6.16E−26**	**1.51E−25**	−5.9348
F2	1.19E−10	4.37E−11	**1.59E−16**	**1.02E−16**	−14.9151
F3	7.7917	11.4700	**1.8868**	**2.5894**	−2.7505
F4	0.0075	0.0079	**1.43E−04**	**4.51E−04**	−5.0925
F5	**61.7308**	**0.8211**	61.8402	0.8206	0.5162
F6	**4.4760**	**0.8275**	6.8053	0.7593	11.3600
F7	0.0047	0.0020	**0.0020**	**9.49E−04**	−6.6804
F8	−10,750.24	2597.67	**−6091.45**	**2222.02**	−7.4647
F9	5.8241	5.9341	**3.79E−14**	**6.89E−14**	−5.3757
F10	9.81E−10	4.27E−10	**9.53E−14**	**2.17E−14**	−12.5823
F11	0.0043	0.0081	**8.92E−04**	**0.0049**	−1.9718
F12	**0.1614**	**0.0429**	0.3101	0.0811	8.8772
F13	**3.3001**	**0.3122**	3.8761	0.2947	7.3485

respectively and we can conclude that the performance of the proposed method improved significantly in comparison with 30 dimensions because it is better than the original method in 9 of the 13 benchmarks that were analyzed.

In addition we are presenting the results when the problems have 128 dimensions respectively and with the help of Table 4.7 we can say that for this number of dimensions, the randomness based on sine waves is better than the simple random numbers in 8 of the 13 benchmark functions according to the hypothesis test presented above.

In the GWO algorithm we also are presenting the results when the parameters of the metaheuristic change, in order to test the performance sensitivity of the proposed method. The number of individuals, iterations and dimensions are 30, 500 and 64 respectively. Experiments were independently executed 51 times and were tested with the first 13 benchmark functions that were presented in this book, and finally the parameters that we modified for the experiments are the following.

In this algorithm, we present two variations, the first variation has the following values in the parameters, "a" parameter: 1.5, and "C" parameter: 1.5. In Table 4.8 we can find the statistical results for the 13 benchmark functions that were analyzed.

Table 4.9 shows the hypothesis test results between the original randomness method and the proposed randomness with the second variant in the algorithm that is: "a" parameter: 2.5, and "C" parameter: 2.5

Finally in Fig. 4.8 we are illustrating in a graphical way the results of the hypothesis tests, more specifically the z-values and the features are the same as mentioned above, as a brief explanation, while the blue and green bars are less than −1.645 the proposed method is better than the conventional randomness method.

According to Tables 4.8 and 4.9 and with Fig. 4.8 we can note that for the variation 1 the proposed method is better than the original method in 12 of the 13 benchmark

Table 4.7 Comparison between the original and proposed method in the GWO with 128 Dimensions

128 Dimensions					
Function	GWO	STD	GWO-SW	STD	Z value
F1	1.31E−10	9.97E−11	**3.54E−15**	**5.04E−15**	−7.1758
F2	5.94E−07	1.88E−07	**4.07E−10**	**1.84E−10**	−17.2787
F3	3177.59	3313.29	**318.4972**	**559.1867**	−4.6605
F4	**4.7011**	**3.3964**	5.2794	5.0783	0.5185
F5	**126.00**	**0.5279**	126.0598	0.5786	0.3969
F6	**15.1484**	**1.2651**	19.1919	1.0172	13.6430
F7	0.0090	0.0035	**0.0043**	**0.0022**	−6.2432
F8	−19,857.13	3370.99	**−9693.91**	**4955.85**	−9.2875
F9	15.2971	10.4893	**1.27E−10**	**6.53E−10**	−7.9877
F10	1.06E−06	3.51E−07	**4.90E−09**	**2.92E−09**	−16.4843
F11	5.09E−03	0.0120	**1.76E−15**	**1.60E−15**	−2.3239
F12	**0.3796**	**0.0630**	0.5259	0.0733	8.2988
F13	**9.5320**	**0.4411**	10.0525	0.3823	4.8836

Table 4.8 Comparison between the original and proposed method in the GWO algorithm with 64 Dimensions and variation 1

64 Dimensions					
Function	GWO	STD	GWO-SW	STD	Value of Z
F1	3.30E−09	1.71E−09	**1.24E−29**	**1.84E−29**	−13.8269
F2	5.79E−06	1.83E−06	**7.01E−18**	**5.37E−18**	−22.5763
F3	49.1258	46.9403	**1.3447**	**8.7651**	−7.1458
F4	0.2283	0.1403	**5.26E−07**	**1.05E−06**	−11.6232
F5	**61.9902**	**0.8130**	61.8012	0.7533	−1.2176
F6	7.4625	0.9797	**6.5199**	**0.7669**	−5.4103
F7	0.0150	0.0049	**0.0019**	**0.0012**	−18.5302
F8	−10,995.20	1412.86	**−7806.75**	**2441.06**	−8.0732
F9	72.4231	19.2216	**4.46E−14**	**6.85E−14**	−26.9075
F10	9.25E−06	2.86E−06	**5.18E−14**	**9.79E−15**	−23.1062
F11	0.0119	0.0141	**1.51E−04**	**0.0011**	−5.9635
F12	0.4837	0.1991	**0.2962**	**0.1024**	−5.9803
F13	4.8219	0.4567	**3.9713**	**0.3385**	−10.6864

Table 4.9 Comparison between the original and proposed method in the GWO algorithm with 64 Dimensions and variation 2

64 Dimensions

Function	GWO	STD	GWO-SW	STD	Value of Z
F1	2.14E−15	1.52E−15	**4.04E−18**	**7.38E−18**	−10.0183
F2	5.82E−10	3.28E−10	**5.93E−12**	**6.01E−12**	−12.5334
F3	91.4090	190.4643	**4 0.6136**	**77.7996**	−1.7631
F4	0.0765	0.2436	**0.0074**	**0.0125**	−2.0222
F5	**61.7871**	**0.7350**	61.9672	0.5588	1.3934
F6	**4.6424**	**0.7323**	7.1327	0.6375	18.3156
F7	**0.0041**	**0.0020**	0.0035	0.0022	−1.4305
F8	−8343.76	3665.59	**−5912.95**	**487.20**	−4.6945
F9	1.5618	2.9193	**1.76E−12**	**9.51E−12**	−3.8207
F10	**0.8122**	**4.0604**	16.2164	8.4229	11.7649
F11	0.0032	0.0082	**0.0007**	**0.0049**	−1.8586
F12	**0.1933**	**0.0849**	0.3303	0.0915	7.8414
F13	**3.4078**	**0.3735**	4.0903	0.3194	9.9168

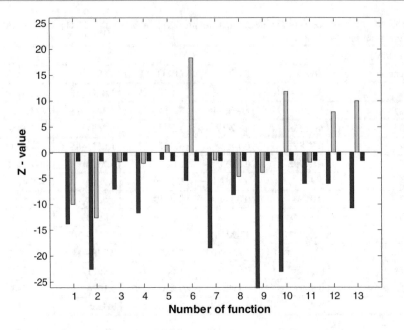

Fig. 4.8 Comparison between original GWO and the proposed method in both variants

functions that were analyzed and for the modification into variation 2, the Sine Waves randomness has better performance in 7 of the 13 benchmark functions that were mentioned above.

According to the variations and the statistical results that we presented above, we can note that for the GWO algorithm the proposed method has a good performance even when the parameters in the algorithm change its values.

4.3.1.2 Comparison with Second Case Study

In addition, we tested both methods (original and proposed method) in the GWO algorithm with the benchmark functions that were presented in the CEC 2015 competition with 30 and 50 dimensions respectively in order to show the performance of the proposed method in comparison with the traditional method.

Table 4.10 summarizes the results for the case of 30 dimensions and according to the statistical study we can say that the new method is better than the conventional method in 6 of the 15 benchmark functions that were analyzed in the CEC 2015 competition. On the other hand we can find in Table 4.11 the results of when the problems have 50 dimensions as complexity and we can conclude that for this number of dimensions the proposed method improved the results in 8 of the 15 benchmark functions in comparison the original method of randomness.

Table 4.10 Comparison between the original and proposed method in the GWO with 30 Dimensions for CEC 15 benchmark functions

30 Dimensions					
Function	GWO	STD	GWO-SW	STD	Z-Value
F1	2.98E+07	2.11E+07	**2.18E+07**	**1.24E+07**	−2.3369
F2	**2.41E+09**	**1.79E+09**	2.44E+09	1.40E+09	0.1044
F3	**20.9484**	**4.93E−02**	20.9436	5.71E−02	−0.4489
F4	**101.6677**	**2.83E+01**	107.2801	4.23E+01	0.7873
F5	**2917.71**	**9.41E+02**	4102.99	1.72E+03	4.3086
F6	1.38E+06	1.24E+06	**9.75E+05**	**7.91E+05**	−1.9617
F7	21.2706	8.63E+00	**18.3992**	**4.47E+00**	−2.1098
F8	**2.66E+05**	**2.41E+05**	3.47E+05	2.73E+05	1.5922
F9	138.1955	6.59E+01	**118.2747**	**4.18E+01**	−1.8229
F10	**1.50E+06**	**1.10E+06**	1.64E+06	1.08E+06	0.6321
F11	**776.7721**	**1.03E+02**	758.3066	8.97E+01	−0.9649
F12	**117.6998**	**3.08E+01**	117.6216	3.06E+01	−0.0129
F13	0.0712	3.16E−02	**0.0365**	**8.39E−03**	−7.5746
F14	**3.58E+04**	**2.67E+03**	3.53E+04	1.41E+03	−1.0636
F15	180.8290	2.07E+02	**123.2268**	**1.30E+01**	−1.9844

Table 4.11 Comparison between the original and proposed method in the GWO with 50 Dimensions for CEC 15 benchmark functions

50 Dimensions

Function	GWO	STD	GWO-SW	STD	Z-Value
F1	**8.95E+07**	**7.13E+07**	7.45E+07	4.25E+07	−1.2917
F2	1.01E+10	4.83E+09	**8.07E+09**	**3.35E+09**	−2.5068
F3	**21.1301**	**4.21E−02**	21.1393	3.93E−02	1.1430
F4	**220.8364**	**3.47E+01**	225.6384	5.26E+01	0.5437
F5	**5253.02**	**7.31E+02**	7456.82	3.27E+03	4.6957
F6	5.02E+06	3.68E+06	**3.32E+06**	**1.59E+06**	−3.0273
F7	109.2351	3.00E+01	**82.9092**	**2.63E+01**	−4.7113
F8	**2.53E+06**	**1.37E+06**	2.30E+06	1.64E+06	−0.7625
F9	183.7510	1.14E+02	**151.0719**	**6.69E+01**	−1.7690
F10	6.36E+06	4.86E+06	**3.76E+06**	**2.43E+06**	−3.4209
F11	**1184.1528**	**1.13E+02**	1188.8295	8.80E+01	0.2331
F12	**177.2905**	**3.86E+01**	186.4990	3.26E+01	1.3010
F13	0.4364	2.27E−01	**0.2322**	**7.47E−02**	−6.1076
F14	8.06E+04	4.40E+03	**7.28E+04**	**6.50E+03**	−7.1515
F15	443.3092	3.52E+02	**226.6245**	**1.71E+02**	−3.9578

Based on the hypothesis test in Tables 4.10 and 4.11; we can note that there is enough evidence to state that the proposed method improved the results with more complex problems in the GWO algorithm.

4.3.1.3 Comparison with the Third Case Study

In this case, with the GWO algorithm we are also presenting the results obtained with the original method and the proposed method for the CEC 17 Competition on the constrained functions. In order to respect the rules of this competition, we used 30 individuals and 33,333 iterations for 50 dimensions and the configuration of the parameters are the ones that the original paper suggested. The number of penalties in each function in both cases is similar; the main change is in the optimal value and standard deviation that has the algorithm with different randomness method.

Table 4.12 and Fig. 4.9 show the averages, standard deviations and the hypothesis test results for the first 10 problems with constraints in CEC 17 competition in 31 independent executions.

Figure 4.9 shows the results of the hypothesis tests in a graphical way and according these results, we can mention that the red bars represent the problems where the proposed method has a poor performance than the original method and the green bars are the results where the proposed method has better performance than the original method.

Table 4.12 Comparison between the original and proposed method in the GWO with 50 Dimensions for CEC'17 constrained functions

50 Dimensions					
Function	GWO	STD	GWO-SW	STD	Z-Value
F1	8657.98	2032.415	**7752.499**	**2254.179**	−2.1305
F2	8657.98	2032.42	**7647.669**	**1664.223**	−2.7467
F3	9734.57	2605.37	**9014.268**	**1050.921**	−1.8310
F4	**297.0701**	**40.1308**	336.0141	68.0900	3.5188
F5	**31,557.05**	**23,287.72**	38,044.93	21,670.60	1.4565
F6	**513.4068**	**136.0203**	529.6185	87.5052	0.7158
F7	−1450.85	238.30	**−1765.58**	**136.99**	−8.1771
F8	−45.9427	40.9837	**−90.3668**	**0.00**	−7.7409
F9	−53.0457	41.4657	**−90.0593**	**5.78E−14**	−6.3747
F10	−28.3182	2.2731	**−42.4080**	**4.9116**	−18.5919

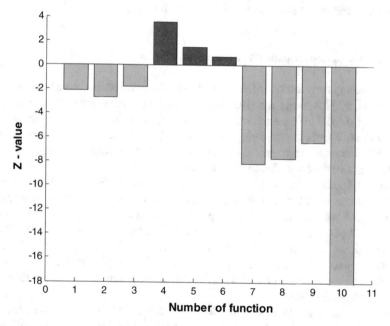

Fig. 4.9 Comparison between original and the proposed method in GWO algorithm for CEC 17 constrained problems

Based on Table 4.12 and Fig. 4.9 we can conclude that the proposed method improves the results of the GWO algorithm in 7 of the 10 functions that were analyzed (CEC 17).

4.3.2 Adaptation of the Firefly Algorithm

In the firefly algorithm [18] the randomness is represented with a Gaussian or uniform distribution [19] multiplied with an alpha factor as we can note in Eq. 4.7 at the final term.

$$x_i^{t+1} = x_i^t + \beta_0 e^{-\gamma r_{ij}^2} \left(x_j^t - x_i^t\right) + \alpha_t \epsilon_i^t \qquad (4.7)$$

In this algorithm we are replacing the alpha factor α_t for the amplitude and the uniform distribution ϵ_i^t for the wave values, so the Equation that is representing the proposed method into the algorithm is the following:

$$x_i^{t+1} = x_i^t + \beta_0 e^{-\gamma r_{ij}^2} \left(x_j^t - x_i^t\right) + |S| \qquad (4.8)$$

It is important to mention that the FA only uses the positive numbers and this is why in the sine wave we are applying the absolute values, in order to respect the range of values that is used in the original algorithm $[0, \alpha_t]$.

4.3.2.1 Comparison with First Case Study

The parameters that we used in this algorithm for the first case study are the following: 20 fireflies, 500 iterations, with the alpha parameter in decrement with respect to the iterations from 0.8 to 0, the minimum value of beta was 0.2 and for the gamma value we considered 1, these parameters were used based on the paper of Lagunes et al. [18]. It is important to mention that the 51 independent executions have the same parameters and the only change is in the randomness method.

Table 4.13 shows the Z-Values as a result of hypothesis tests between the FA (with randomness as Gauss distribution) and FA-SW (with randomness of proposed method) and we can notice that the proposed method is better than the original method in 7 of the 13 conventional benchmark functions that were described above. In addition we can find in Table 4.14 the Z-values when the problems have 50 dimensions and for this case we can conclude that also the proposed method is better than the original method in 7 of the 13 functions that were analyzed but changing the benchmark functions.

In addition, we are presenting the results of the first set of benchmark functions with other values in the parameters that are critical in the Firefly Algorithm, in order to show the performance of the proposed method when the parameters are different. It is important to mention that we used 50 dimensions, 20 fireflies and 500 iterations in each experiment, the results are the averages and standard deviations of 51 independent executions, and the new parameters and results are the following.

The first variation in the algorithm is the following: alpha: 0.5, beta: 0.5 and gamma 1 respectively. Table 4.15 shows the statistical results of the 13 benchmark functions that were analyzed in this book.

Table 4.13 Comparison between the original and proposed method in the FA with 30 Dimensions

30 Dimensions

Function	FA	STD	FA-SW	STD	Z-Value
F1	0.0194	0.0050	**0.0153**	**0.0065**	−3.5353
F2	**0.4554**	**0.1543**	0.4284	0.1289	−0.9496
F3	**3886.15**	**1366.59**	3551.44	1216.29	−1.2937
F4	**0.1176**	**0.0325**	0.1094	0.0268	−1.3765
F5	**222.6977**	**397.6304**	203.4445	494.3338	−0.2146
F6	0.0216	0.0062	**0.0157**	**0.0057**	−4.9536
F7	**0.0643**	**0.0375**	0.0619	0.0431	−0.2971
F8	−7057.79	770.29	**−6724.17**	**707.52**	−2.2555
F9	49.4399	17.1045	**42.8881**	**11.3371**	−2.2576
F10	0.0780	0.0551	**0.0586**	**0.0147**	−2.4055
F11	0.0104	0.0035	**0.0085**	**0.0033**	−2.7929
F12	**4.85E−04**	**3.36E−04**	0.0025	0.0148	0.9625
F13	0.0049	0.0039	**0.0033**	**0.0016**	−2.6839

Table 4.14 Comparison between the original and proposed method in the FA with 50 Dimensions

50 Dimensions

Function	FA	STD	FA-SW	STD	Z-Value
F1	0.0921	0.0210	**0.0731**	**0.0207**	−4.5562
F2	**2.0574**	**1.8178**	2.2224	1.6003	0.4818
F3	18,803.18	3879.59	**16,827.89**	**3573.77**	−2.6480
F4	14.6682	7.3989	**7.7247**	**6.1767**	−5.0941
F5	**299.2605**	**554.2157**	252.1285	382.4818	−0.4949
F6	0.0902	0.0194	**0.0740**	**0.0199**	−4.1218
F7	**0.1203**	**0.0415**	0.1220	0.0425	0.2024
F8	−11,288.36	1016.31	**−10,446.64**	**844.82**	−4.5036
F9	**93.8296**	**25.4291**	100.2590	22.6229	1.3357
F10	0.1949	0.0604	**0.1635**	**0.0398**	−3.0695
F11	0.0284	0.0080	**0.0207**	**0.0051**	−5.7389
F12	**0.0443**	**0.0954**	0.0372	0.1111	−0.3428
F13	**0.0165**	**0.0065**	0.0155	0.0055	−0.8305

The second variation in the algorithm was: alpha: 0.7, beta: 0.8 and gamma 0.8 respectively. Table 4.16 shows the statistical results of the 13 benchmark functions that were used to test the proposed method.

Finally in Fig. 4.10 we are illustrating in a graphical way the results of the hypothesis tests, more specifically the z-values. In this case the blue bar represents the z

Table 4.15 Comparison between the original and proposed method in the FA with 50 Dimensions and variation 1

50 Dimensions

Function	FA	STD	FA-SW	STD	Z-Value
F1	**0.0977**	**0.0292**	0.0924	0.0320	−0.8797
F2	6.1172	4.3566	**2.6222**	**1.6314**	−5.3652
F3	16,684.99	3713.66	**10,771.08**	**2413.43**	−9.5358
F4	16.2072	4.8708	**9.6982**	**2.6409**	−8.3895
F5	672.0072	1067.5816	**351.8132**	**476.1820**	−1.9561
F6	0.1011	0.0253	**0.0856**	**0.0202**	−3.4117
F7	0.1688	0.0627	**0.1433**	**0.0485**	−2.2990
F8	**−9897.64**	1290.47	−9798.69	1530.95	−0.3529
F9	116.4533	27.5726	**97.9171**	**19.3021**	**−3.9330**
F10	**0.4116**	**0.2692**	0.4106	0.2257	−0.0200
F11	**0.0370**	**0.0094**	0.0334	0.0123	−1.6301
F12	**1.3505**	**1.1046**	1.1275	1.0892	−1.0267
F13	0.8113	2.2687	**0.2473**	**0.8867**	−1.6534

Table 4.16 Comparison between original and proposed method in the FA with 50 Dimensions and variation 2

50 Dimensions

Function	FA	STD	FA-SW	STD	Z-Value
F1	3.5472	4.9279	**1.6513**	**1.8790**	−2.5671
F2	**23.5032**	**11.2461**	24.6619	17.1214	0.4040
F3	21,470.97	4372.85	**19,555.45**	**4415.63**	−2.2012
F4	41.8120	7.6392	**34.6967**	**6.6137**	−5.0288
F5	6113.53	10,216.69	**3375.71**	**1020.58**	−1.9042
F6	**2.3755**	**3.4473**	1.8711	2.3143	−0.8676
F7	**0.8897**	**0.3236**	0.9987	0.5129	1.2827
F8	10,573.29	1049.12	**10,283.63**	**550.44**	−1.7460
F9	**161.5472**	**31.7380**	162.7472	28.9984	0.1993
F10	2.2555	0.6345	**2.0838**	**0.2364**	−1.8106
F11	0.5132	0.3037	**0.4111**	**0.2523**	−1.8478
F12	7.2384	2.4273	**6.6715**	**1.2896**	−1.4730
F13	45.1570	26.3890	52.1288	22.5756	1.4337

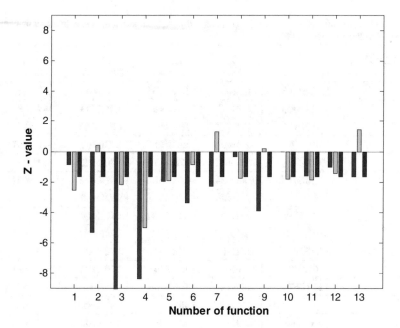

Fig. 4.10 Comparison between original FA and the proposed method in both variants

value of the hypothesis test with variation 1, the green bar represents the z values of the hypothesis with variation 2 and finally the red bar represents the critical value according of the hypothesis test that was described above, which in this case is −1.645. Therefore, while the blue and green bars are less than −1.645 the proposed method is better.

According to Tables 4.15 and 4.16 and with the help of Fig. 4.10 we can conclude that for the variation 1, the proposed method is better than the conventional method in 8 of the 13 benchmark functions that were analyzed and for the variation 2 in the parameters, the Sine Waves method has better performance also in 8 of the 13 benchmark functions that were mentioned above.

According with the variations and the statistical results that we presented above; we can note that for the firefly algorithm the proposed method has a good performance even when the parameters in the algorithm change its values.

4.3.2.2 Comparison with Second Case Study

In this metaheuristic we also present the performance of the proposed method when the problems are more complex, and in this case we only changed the number of iterations and the Fireflies in order to satisfy the maximum number of function evaluations allowed in the competition. Table 4.17 shows the results for the 15 benchmark functions that were presented in the CEC'15 competition with 30 dimensions respectively

Table 4.17 Comparison between the original and proposed method in the FA with 30 Dimensions for CEC'15 benchmark functions

30 Dimensions					
Function	FA	STD	FA-SW	STD	Z-Value
F1	2.37E+06	1.02E+06	**1.16E+06**	**5.08E+05**	−7.5859
F2	8.19E+03	5.09E+03	**4.11E+03**	**3.31E+03**	−4.7883
F3	20.0197	2.71E−02	**20.0024**	**1.59E−03**	−4.5586
F4	88.9176	2.22E+01	**32.4259**	**9.23E+00**	−16.7757
F5	3215.76	6.27E+02	**1721.72**	**6.40E+02**	−11.9057
F6	1.03E+05	6.68E+04	**5.11E+04**	**3.31E+04**	−4.9602
F7	15.5413	2.13E+00	**9.3701**	**1.62E+00**	−16.4516
F8	6.64E+04	3.17E+04	**3.80E+04**	**2.17E+04**	−5.2720
F9	102.8650	2.75E−01	**102.5117**	**2.20E−01**	−7.1646
F10	1.15E+05	6.87E+04	**7.51E+04**	**3.72E+04**	−3.6102
F11	608.8026	2.55E+02	**427.7404**	**2.29E+01**	−5.0581
F12	**110.8674**	**1.83E+01**	106.2308	1.35E+01	−1.4578
F13	0.0441	1.07E−02	**0.0290**	**2.17E−03**	−9.8681
F14	3.41E+04	1.25E+03	**3.17E+04**	**8.91E+02**	−11.2439
F15	100.0136	1.15E−03	**100.0097**	**9.36E−04**	−18.5192

and according to the hypothesis test, the proposed method has better performance than the conventional method in 14 of the 15 functions of the CEC'15 competition.

In Table 4.18 we can find the z-values for the hypothesis test in the CEC'15 benchmark functions among both methods, but in this case with 50 dimensions respectively and we can notice that the proposed randomness is better than the uniform distribution for generating randomness in 13 of the 15 benchmark functions that were analyzed.

4.3.2.3   Comparison for the Third Case Study

In this case, for the FA algorithm we are also presenting the results obtained with the original method (Gaussian) and the proposed method (Sine wave) for the constrained functions of the CEC 17 Competition. In order to satisfy the rules of this competition, we used 30 individuals and 33,333 iterations for 50 dimensions and the configuration of the parameters are the ones suggested in the original paper. The number of penalties for each function, in both cases is similar; the main change is in the optimal value and standard deviation that the algorithm has with different randomness methods. Table 4.19 shows the averages, standard deviations and the hypothesis test results for the first 10 problems with constraints in the CEC 17 competition in 31 independent executions.

Table 4.18 Comparison between the original and proposed method in the FA with 50 Dimensions for CEC'15 benchmark functions

50 Dimensions					
Function	FA	STD	FA-SW	STD	Z-Value
F1	1.16E+07	4.08E+06	**4.11E+06**	**1.35E+06**	−12.3876
F2	2.14E+04	1.61E+04	**1.05E+04**	**1.03E+04**	−4.0605
F3	20.0167	2.26E−02	**20.0036**	**1.37E−03**	−4.1521
F4	206.0428	4.31E+01	**66.8798**	**1.47E+01**	−21.8368
F5	5938.56	7.74E+02	**3002.40**	**6.91E+02**	−20.2168
F6	5.11E+05	2.79E+05	**2.35E+05**	**1.29E+05**	−6.4071
F7	57.4436	1.19E+01	**47.0772**	**2.22E+00**	−6.1020
F8	3.40E+05	1.62E+05	**1.55E+05**	**7.38E+04**	−7.4252
F9	104.7630	3.65E−01	**104.0353**	**2.10E−01**	−12.3556
F10	2.33E+05	1.49E+05	**1.31E+05**	**6.95E+04**	−4.4233
F11	1228.9662	1.25E+02	**444.5958**	**1.94E+01**	−44.2072
F12	195.1389	2.09E+01	**167.1883**	**4.53E+01**	−4.0030
F13	0.2036	6.06E−02	**0.0951**	**6.37E−03**	−12.7251
F14	**6.65E+04**	**7.48E+03**	6.44E+04	6.43E+03	−1.5428
F15	**100.0362**	**1.67E−01**	100.0091	7.78E−04	−1.1580

Table 4.19 Comparison between the original and proposed method in the FA with 50 Dimensions for CEC'17 constrained functions

50 Dimensions					
Function	FA	STD	FA-SW	STD	Z-Value
F1	0.0564	0.0142	**0.0295**	**0.0060**	−12.4664
F2	0.0563	0.0129	**0.0307**	**0.0080**	−12.0081
F3	0.0585	0.0133	**0.0273**	**0.0060**	−15.2299
F4	75.4605	13.3594	**70.2809**	**15.0758**	−1.8363
F5	**83.0312**	**96.7234**	489.7094	1204.31	2.4038
F6	78.2756	21.2956	**72.2335**	**10.5380**	−1.8160
F7	**−2852.58**	**201.07**	−2352.80	173.44	13.4410
F8	−90.36682	0.00	−90.36682	0.00	−
F9	−90.0593	5.78E−14	−90.0593	5.78E−14	0.0000
F10	−53.1833	4.33E−14	−53.1833	4.33E−14	0.0000

Table 4.19 shows interesting results for the hypothesis test, according to the performance of the first 7 functions we can conclude that the proposed method is better than the uniform distribution in 5 of the 7 functions that were analyzed. In addition for the 3 last problems that were analyzed, the results are exactly the same, in addition for the function number 8 the FA algorithm has exactly the same values in all of their

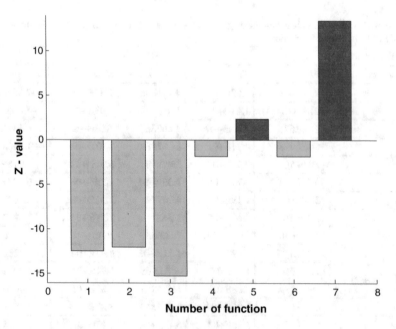

Fig. 4.11 Comparison between original and the proposed method in FA algorithm for CEC 17 constrained problems

results (31 independent executions), and this the reason of the value zero in standard deviation in function number 8.

Figure 4.11 shows the results of the hypothesis tests in a graphical way of the first 7 problems that were analyzed in the CEC 17 constrained problems competition and according to these results, we can mention that the red bars represent the problems, where the proposed method has a poor performance in respect to the original method and the green bars are the results where the proposed method has better performance than the original method.

4.3.3 Adaptation of Flower Pollination Optimization

In this algorithm [20] the randomness factor is represented as a Levy flight [21] in the global pollination as described previously. We can note this parameter as L in Eq. 4.9.

$$x_i^{t+1} = x_i^t + L\left(x_i^t - g_*\right)$$
(4.9)

In this case we are replacing the original randomness by adding the proposed method as shown in the following Equation:

$$x_i^{t+1} = x_i^t + S(x_i^t - g_*)$$
(4.10)

where S is a vector of randomness according to the sine wave method, and for this book the values of the Levy flights have a range from $[-1, 1]$ so the amplitude of the sine wave in this algorithm is 1.

4.3.3.1 Comparison with First Case Study

The features for testing the algorithm with the first case study are: 20 flowers, the probability of the two different types of pollination (global and local) is 0.8 as mentioned in the original paper [12], and finally the number of iterations is 2000.

In Table 4.20 we can find the results obtained between the original method (in this case Levy flights and it is denoted as FPA) and the proposed method (denoted as FPA-SW), with the first set of benchmark functions and 30 dimensions. We can conclude that for this number of dimensions, the proposed method is better than the original method in 7 of the 13 benchmark functions that were analyzed in this book (we show in bold the best results).

In Table 4.21 we can find the results of the hypothesis test when the set of problems have 50 dimensions respectively and we can conclude that for this number of dimensions, the randomness based on the sine wave has better performance than the original randomness method in 6 of the 13 benchmark functions that were analyzed in this book.

Table 4.20 Comparison between the original and proposed method in the FPA with 30 Dimensions

30 Dimensions					
Function	FPA	STD	FPA-SW	STD	Z-Value
F1	537.6191	272.9331	**26.5576**	**26.9570**	−13.1763
F2	5.9401	1.9507	**2.4858**	**2.7056**	−7.3229
F3	**405.5312**	**202.6527**	1234.4592	623.3695	8.9421
F4	**15.4369**	**2.7773**	25.6416	5.0315	12.5556
F5	74,306.40	65,916.21	**9997.58**	**10,250.64**	−6.8167
F6	501.8553	248.1471	**32.6734**	**37.4936**	−13.2195
F7	**0.1664**	**0.1026**	0.2449	0.1905	2.5657
F8	−7837.470	227.277	**−6302.802**	1202.992	−8.8638
F9	**72.1825**	**13.0220**	124.5127	53.0119	6.7786
F10	5.0400	1.5558	**4.3322**	**1.1568**	−2.5818
F11	5.7296	2.5920	**1.3819**	**0.6521**	−11.5022
F12	**8.1251**	**14.0604**	120.7631	509.7258	1.5620
F13	**8861.34**	**40,376.08**	14,906.67	47,850.90	0.6828

Table 4.21 Comparison between the original and proposed method in the FPA with 50 Dimensions

50 Dimensions					
Function	FPA	STD	FPA-SW	STD	Z-Value
F1	2161.39	722.8734	**1385.79**	**622.2149**	−5.7501
F2	20.3239	3.7079	**11.3151**	**2.8555**	−13.6115
F3	**1648.40**	**723.4265**	4458.31	1834.35	10.0764
F4	**20.5684**	**3.0223**	32.5373	4.1760	16.4181
F5	5.07E+05	3.28E+05	**3.37E+05**	**2.24E+05**	−3.0250
F6	2160.26	584.0145	**1473.85**	**628.6582**	−5.6565
F7	**0.9261**	**0.4351**	1.6387	0.9545	4.8041
F8	**−1.15E+04**	**389.3387**	−1.13E+04	2308.40	−0.6927
F9	162.6066	17.9314	**139.2513**	**51.3433**	−3.0367
F10	**5.1718**	**1.4852**	5.7776	1.9440	1.7507
F11	21.3003	6.5287	**14.6523**	**5.1879**	−5.6371
F12	**68.4205**	**369.3284**	5341.2459	10,674.6636	3.4907
F13	**2.15E+05**	**4.97E+05**	3.66E+05	4.26E+05	1.6350

4.4 String Theory Algorithm

In the following section, we are presenting the results and comparison of the String Theory Algorithm with three different algorithms that are GWO, FA and FPA respectively, with a set of traditional benchmark functions and the benchmark functions that were presented in the CEC 15 competition. Finally, we presented the String Theory Algorithm in the fuzzy logic problem that we describe in detail above.

4.4.1 Comparison with Grey Wolf Optimizer

The first algorithm that we use in order to compare the String Theory Algorithm is the Grey Wolf Optimizer with the first and second case study.

4.4.1.1 First Case Study

In this work, we realized a comparison with respect to the Grey Wolf Algorithm (GWO) with the first case study of this work. The parameters that we are using for this comparison in the GWO algorithm are the ones proposed by the original author of the paper [22]. Table 4.22 shows the hypothesis test results between the GWO and STA metaheuristics and we can note an interesting behavior and is that although the averages and standard deviations of the proposed method are better than the GWO

Table 4.22 Comparison between GWO and STA with traditional benchmark functions and 30 dimensions

30 Dimensions					
Function	GWO	STD	STA	STD	Z-Value
F1	**6.59E−28**	**6.34E−05**	2.93E−38	1.34E−37	−7.42E−23
F2	**7.18E−17**	**0.0290**	1.46E−24	4.67E−24	−1.77E−14
F3	**3.29E−06**	**79.1496**	6.37E−08	4.43E−07	−2.91E−07
F4	**5.61E−07**	**1.3151**	6.46E−06	3.53E−05	3.20E−05
F5	**26.8126**	**69.9050**	28.2172	0.5788	0.1435
F6	**0.8166**	**1.26E−04**	4.4536	0.4451	58.3535
F7	**0.0022**	**0.1003**	9.12E−04	8.78E−04	−0.0927
F8	**−6123.10**	**4087.44**	−5431.064	671.539	−1.1931
F9	**0.3105**	**47.3561**	0	0	−0.0468
F10	**1.06E−13**	**0.0778**	7.78E−15	1.93E−15	−9.01E−12
F11	0.0045	0.0067	**0**	**0**	−4.8099
F12	**0.0534**	**0.0207**	0.5002	0.2159	14.7107
F13	**0.6545**	**0.0045**	2.2651	0.1961	58.6442

Table 4.23 Comparison between GWO and STA with traditional benchmark functions and 50 dimensions

64 Dimensions					
Function	GWO	STD	STA	STD	Z-Value
F1	8.56E−17	7.90E−17	**3.69E−24**	**7.62E−24**	−7.7368
F2	1.19E−10	4.37E−11	**6.86E−17**	**1.39E−16**	−19.3485
F3	7.7917	11.4700	**0.0918**	**0.3970**	−4.7912
F4	**0.0075**	**0.0079**	37.4189	18.9449	14.1025
F5	**61.7308**	**0.8211**	62.3561	0.4758	4.7055
F6	**4.4760**	**0.8275**	11.7088	0.5860	50.9375
F7	0.0047	0.0020	**0.0016**	**0.0015**	−8.8226
F8	−10,750.24	2597.67	**−9759.06**	**1184.45**	−2.4793
F9	5.8241	5.9341	**8.92E−15**	**3.09E−14**	−7.0091
F10	9.81E−10	4.27E−10	**1.80E−13**	**2.33E−13**	−16.4020
F11	0.0043	0.0081	**2.05E−16**	**1.38E−15**	−3.8158
F12	**0.1614**	**0.0429**	0.6906	0.1006	34.5663
F13	**3.3001**	**0.3122**	5.5964	0.1738	45.8953

algorithm there is not enough evidence to prove that STA has better performance according to the hypothesis test that we described above.

In Table 4.23 we can note the comparison between the GWO and STA metaheuristics when the first set of problems has 64 dimensions. In this case, we can note that the STA metaheuristic has better performance than the GWO algorithm in 8 of the 13 benchmark functions that we analyzed in this work. Finally, as a brief conclusion, we can note that the proposed method improves the results when the problems are more complex according to the hypothesis test that we have presented in this work.

4.4.1.2 Second Case Study

On the other hand, we are presenting in Table 4.24 the hypothesis test results between the GWO and STA metaheuristics with the Benchmark Functions of the CEC 15 Competition. The hypothesis test results show that for 30 dimensions, the proposed method has better performance in 10 of the 15 benchmark functions that were analyzed in this work.

Finally, for this case study, in Table 4.25 we can find the z-values for the hypothesis tests when the problems have 50 dimensions for the problems of CEC 15 competition. The comparison is between the GWO and STA metaheuristics and we can note that for these dimensions the proposed method is better than the GWO algorithm in 11 of the 15 benchmark functions that were proposed in this work.

Table 4.24 Comparison between GWO and STA algorithms with 30 Dimensions for CEC'15 benchmark functions

30 Dimensions					
Function	GWO	STD	STA	STD	Z-Value
F1	2.98E+07	2.11E+07	**1.29E+06**	**6.27E+05**	−9.6446
F2	2.41E+09	1.79E+09	**8.55E+03**	**5.79E+03**	−9.6427
F3	20.9484	4.93E−02	**20.0082**	**1.95E−02**	−126.6638
F4	**101.6677**	**2.83E+01**	171.3461	3.53E+01	10.9888
F5	**2917.71**	**9.41E+02**	3103.68	4.82E+02	1.2566
F6	1.38E+06	1.24E+06	**6.56E+04**	**3.11E+04**	−7.5401
F7	21.2706	8.63E+00	**14.8285**	**9.51E+00**	−3.5821
F8	2.66E+05	2.41E+05	**3.19E+04**	**2.08E+04**	−6.9243
F9	**138.1955**	**6.59E+01**	120.6151	6.26E+01	−1.3815
F10	1.50E+06	1.10E+06	**2.09E+04**	**9.60E+03**	−9.5816
F11	**776.7721**	**1.03E+02**	875.1474	2.28E+02	2.8044
F12	**117.6998**	**3.08E+01**	131.8211	4.05E+01	1.9815
F13	0.0712	3.16E−02	**0.0297**	**3.83E−03**	−9.3086
F14	3.58E+04	2.67E+03	**3.39E+04**	**1.59E+03**	−4.2774
F15	180.8290	2.07E+02	**100.0084**	**1.32E−03**	−2.7897

Table 4.25 Comparison between GWO and STA algorithms with 50 Dimensions for CEC'15 benchmark functions

50 Dimensions

Function	GWO	STD	STA	STD	Z-Value
F1	8.95E+07	7.13E+07	**3.72E+06**	**1.25E+06**	−8.5864
F2	1.01E+10	4.83E+09	**1.84E+04**	**1.55E+04**	−14.9785
F3	21.1301	4.21E−02	**20.0089**	**9.07E−03**	−186.1177
F4	**220.8364**	**3.47E+01**	357.6976	6.51E+01	13.2413
F5	**5253.02**	**7.31E+02**	5474.37	8.79E+02	1.3829
F6	5.02E+06	3.68E+06	**3.43E+05**	**1.91E+05**	−9.0748
F7	109.2351	3.00E+01	**61.1664**	**2.77E+01**	−8.4155
F8	2.53E+06	1.37E+06	**1.67E+05**	**6.98E+04**	−12.3307
F9	183.7510	1.14E+02	**123.2702**	**9.43E+01**	−2.9239
F10	6.36E+06	4.86E+06	**1.20E+04**	**2.20E+03**	−9.3245
F11	**1184.1528**	**1.13E+02**	1320.9949	1.93E+02	4.3667
F12	**177.2905**	**3.86E+01**	182.6002	3.61E+01	0.7171
F13	0.4364	2.27E−01	**0.1015**	**9.87E−03**	−10.5371
F14	8.06E+04	4.40E+03	**6.44E+04**	**1.06E+04**	−10.0989
F15	443.3092	3.52E+02	**100.0101**	**1.47E−03**	−6.9729

4.4.2 Comparison with the Firefly Algorithm

The second algorithm that we use in order to compare the String Theory Algorithm is the Firefly Algorithm with the first and second case study.

4.4.2.1 First Case Study

In addition, we are presenting a comparison with the Firefly Algorithm [4], and the parameters that we were used for testing the FA with the traditional benchmark functions were analyzed by Lagunes et al. in [18]. Table 4.26 shows the results of the hypothesis test when the set of problems has 30 dimensions and we can note that the proposed method (STA) is better than the FA in 10 of the 13 traditional benchmark functions that we are presenting (Table 4.27).

4.4.2.2 Second Case Study

Table 4.28 shows averages and standard deviations that the STA and FA have respectively with the benchmark functions of CEC 2015 competition and in the last column,

Table 4.26 Comparison between FA and STA with traditional benchmark functions and 30 dimensions

30 Dimensions					
Function	FA	STD	STA	STD	Z-Value
F1	0.0194	0.0050	**2.93E−38**	**1.34E−37**	−28.0091
F2	0.4554	0.1543	**1.46E−24**	**4.67E−24**	−21.0761
F3	3886.15	1366.59	**6.37E−08**	**4.43E−07**	−20.3080
F4	0.1176	0.0325	**6.46E−06**	**3.53E−05**	−25.7980
F5	222.6977	397.6304	**28.2172**	**0.5788**	−3.4929
F6	**0.0216**	**0.0062**	4.4536	0.4451	71.1016
F7	0.0643	0.0375	**9.12E−04**	**8.78E−04**	−12.0772
F8	−7057.79	770.29	**−5431.064**	**671.539**	−11.3680
F9	49.4399	17.1045	**0**	**0**	−20.6420
F10	0.0780	0.0551	**7.78E−15**	**1.93E−15**	−10.1178
F11	0.0104	0.0035	**0**	**0**	−20.9258
F12	**4.85E−04**	**3.36E−04**	0.5002	0.2159	16.5301
F13	**0.0049**	**0.0039**	2.2651	0.1961	82.2996

Table 4.27 Comparison between FA and STA with traditional benchmark functions and 50 dimensions

50 Dimensions					
Function	FA	STD	STA	STD	Z-Value
F1	0.0921	0.0210	**2.82E−28**	**7.49E−28**	−31.3586
F2	2.0574	1.8178	**1.09E−19**	**1.44E−19**	−8.0826
F3	18,803.18	3879.59	**3.87E−04**	**0.0025**	−34.6123
F4	14.6682	7.3989	**6.8004**	**11.7260**	−4.0524
F5	299.2605	554.2157	**48.4362**	**0.4446**	−3.2320
F6	**0.0902**	**0.0194**	8.8501	0.4982	125.4691
F7	0.1203	0.0415	**0.0012**	**0.0011**	−20.5037
F8	−11,288.36	1016.31	**−8208.62**	**997.93**	−15.4414
F9	93.8296	25.4291	**0**	**0**	−26.3508
F10	**0.1949**	**0.0604**	0.4016	2.8682	0.5147
F11	0.0284	0.0080	**2.18E−18**	**1.55E−17**	−25.3846
F12	**0.0443**	**0.0954**	0.6375	0.1449	24.4193
F13	**0.0165**	**0.0065**	4.1973	0.1861	160.3195

Table 4.28 Comparison between FA and STA algorithms with 30 Dimensions for CEC'15 benchmark functions

30 Dimensions

Function	FA	STD	STA	STD	Z-Value
F1	2.37E+06	1.02E+06	**1.29E+06**	**6.27E+05**	−6.4482
F2	**8.19E+03**	**5.09E+03**	8.55E+03	5.79E+03	0.3321
F3	20.0197	2.71E−02	**20.0082**	**1.95E−02**	−2.4583
F4	**88.9176**	**2.22E+01**	171.3461	3.53E+01	14.1088
F5	**3215.76**	**6.27E+02**	3103.68	4.82E+02	−1.0120
F6	1.03E+05	6.68E+04	**6.56E+04**	**3.11E+04**	−3.6163
F7	**15.5413**	**2.13E+00**	14.8285	9.51E+00	−0.5222
F8	6.64E+04	3.17E+04	**3.19E+04**	**2.08E+04**	−6.4961
F9	**102.8650**	**2.75E−01**	120.6151	6.26E+01	2.0258
F10	1.15E+05	6.87E+04	**2.09E+04**	**9.60E+03**	−9.6514
F11	**608.8026**	**2.55E+02**	875.1474	2.28E+02	5.5619
F12	**110.8674**	**1.83E+01**	131.8211	4.05E+01	3.3659
F13	0.0441	1.07E−02	**0.0297**	**3.83E−03**	−9.0380
F14	**3.41E+04**	**1.25E+03**	3.39E+04	1.59E+03	−0.7489
F15	100.0136	1.15E−03	**100.0084**	**1.32E−03**	−20.8314

we can note the hypothesis tests results in order to have a conclusion on which algorithm has better performance in these set of problems. Based on the table we can note that for 30 dimensions the proposed method has better performance in 7 of the 15 problems.

In addition, we have the statistically study in Table 4.29 when the problems have 50 dimensions and in this case, we can find that the STA method is better than the FA only in 8 of the total benchmark problems that are presented.

4.4.3 Comparison with Flower Pollination Algorithm

The third algorithm that we use in order to compare the String Theory Algorithm is the Flower Pollination Algorithm with only the first case study.

4.4.3.1 First Case Study

The comparison that we are presenting is with the Flower Pollination Algorithm [23] with the traditional mathematical problems. We are using the parameters that the author recommended in the original paper. Table 4.30 shows the hypothesis test results when the problems have 30 dimensions, and according to the results we can

Table 4.29 Comparison between FA and STA algorithms with 50 Dimensions for CEC'15 benchmark functions

50 Dimensions					
Function	FA	STD	STA	STD	Z-Value
F1	1.16E+07	4.08E+06	**3.72E+06**	**1.25E+06**	−13.1253
F2	**2.14E+04**	**1.61E+04**	1.84E+04	1.55E+04	−0.9380
F3	20.0167	2.26E−02	**20.0089**	**9.07E−03**	−2.2958
F4	**206.0428**	**4.31E+01**	357.6976	6.51E+01	13.8696
F5	5938.56	7.74E+02	**5474.37**	**8.79E+02**	−2.8317
F6	5.11E+05	2.79E+05	**3.43E+05**	**1.91E+05**	−3.5458
F7	**57.4436**	**1.19E+01**	61.1664	2.77E+01	0.8824
F8	3.40E+05	1.62E+05	**1.67E+05**	**6.98E+04**	−7.0186
F9	**104.7630**	**3.65E−01**	123.2702	9.43E+01	1.4017
F10	2.33E+05	1.49E+05	**1.20E+04**	**2.20E+03**	−10.5567
F11	**1228.9662**	**1.25E+02**	1320.9949	1.93E+02	2.8554
F12	195.1389	2.09E+01	**182.6002**	**3.61E+01**	−2.1438
F13	0.2036	6.06E−02	**0.1015**	**9.87E−03**	−11.8778
F14	**6.65E+04**	**7.48E+03**	6.44E+04	1.06E+04	−1.1866
F15	**100.0362**	**1.67E−01**	100.0101	1.47E−03	−1.1164

Table 4.30 Comparison between FPA and STA with traditional benchmark functions and 30 dimensions

30 Dimensions					
Function	FPA	STD	STA	STD	Z-Value
F1	537.6191	272.9331	**2.93E−38**	**1.34E−37**	−14.0671
F2	5.9401	1.9507	**1.46E−24**	**4.67E−24**	−21.7462
F3	405.5312	202.6527	**6.37E−08**	**4.43E−07**	−14.2908
F4	15.4369	2.7773	**6.46E−06**	**3.53E−05**	−39.6942
F5	74,306.40	65,916.21	**28.2172**	**0.5788**	−8.0474
F6	501.8553	248.1471	**4.4536**	**0.4451**	−14.3147
F7	0.1664	0.1026	**9.12E−04**	**8.78E−04**	−11.5151
F8	−7837.470	227.277	**−5431.064**	**671.539**	−24.2401
F9	72.1825	13.0220	**0**	**0**	−39.5857
F10	5.0400	1.5558	**7.78E−15**	**1.93E−15**	−23.1344
F11	5.7296	2.5920	**0**	**0**	−15.7860
F12	8.1251	14.0604	**0.5002**	**0.2159**	−3.8723
F13	**8861.34**	**40,376.08**	2.2651	0.1961	−1.5669

Table 4.31 Comparison between FPA and STA with traditional benchmark functions and 50 dimensions

50 Dimensions					
Function	FPA	STD	STA	STD	Z-Value
F1	2161.39	722.8734	**2.82E−28**	**7.49E−28**	−21.3529
F2	20.3239	3.7079	**1.09E−19**	**1.44E−19**	−39.1440
F3	1648.40	723.4265	**3.87E−04**	**0.0025**	−16.2725
F4	20.5684	3.0223	**6.8004**	**11.7260**	−8.1196
F5	5.07E+05	3.28E+05	**48.4362**	**0.4446**	−11.0443
F6	2160.26	584.0145	**8.8501**	**0.4982**	−26.3078
F7	0.9261	0.4351	**0.0012**	**0.0011**	−15.1797
F8	−1.15E+04	389.3387	**−8208.62**	**997.93**	−21.8498
F9	162.6066	17.9314	**0**	**0**	−64.7603
F10	5.1718	1.4852	**0.4016**	**2.8682**	−10.5471
F11	21.3003	6.5287	**2.18E−18**	**1.55E−17**	−23.2994
F12	**68.4205**	**369.3284**	0.6375	0.1449	−1.3107
F13	2.15E+05	4.97E+05	**4.1973**	**0.1861**	−3.0900

mention that for this set of problems, the proposed method has better performance in 12 of the 13 problems that were analyzed.

In Table 4.31 we can note that for 50 dimensions the proposed method has better performance also in 12 of the 13 benchmark functions.

4.4.4 Results of Fuzzy Logic Optimization with the Four Case Study

In addition, we are presenting a brief explanation of the methodology to design the fuzzy controller [24]. For the inputs of the fuzzy inference system [25], we are presenting the features that we can find in Fig. 4.12.

where the 1, 2, 10 and 11 points are static points respectively, the 3′, 5′, 7′ and 9′ points are the points that the String Theory Algorithm (STA) searching using their proposed techniques and finally, the 4″,6″ and 8″ points are points that the algorithm generate randomly based on the close front and back point respectively.

In Fig. 4.13 we can find the proposed method for the outputs in the fuzzy system of the mobile robot controller.

In Fig. 4.13 we can notice that the points 1 and 9 are static, and then the points 2′, 4′, 6′ and 8' are the ones that STA will search for in the algorithm and finally, the 3″, 5 ″ and 7 ″ are random points as described above.

In addition, we can mention that 16 points are the total number that the proposed method optimized using the features that we describe in String Theory Algorithm

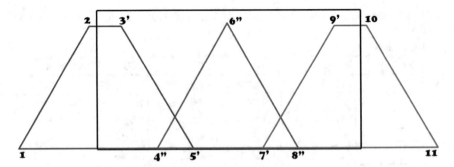

Fig. 4.12 Inputs of fuzzy inference system for representation in the String Theory Algorithm

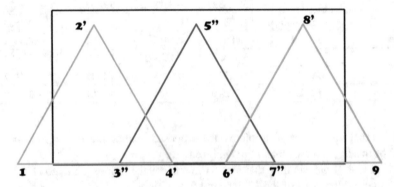

Fig. 4.13 Outputs of fuzzy inference system using String Theory Algorithm

section. Finally, it is important to mention that the goal of this study case is to use a lowest possible computational cost. In this case, we are using only 10,000 evaluation functions in order to find the best structure of the membership functions to solve the autonomous robot mobile problem.

In Fig. 4.14 we can find the best structure of the input membership functions that the String Theory Algorithm finds in 10 independent executions.

In addition in Fig. 4.15, we can note the outputs that were designed with the proposed method for the corresponding autonomous mobile robot problem.

In this problem, it is important to mention the way that we can evaluate the performance of the proposed method in order to solve the autonomous robot problem. In this case, the equation that we used as a fitness value of each String, is the expression of the mean squared error (MSE) [26, 27], in this way measure the efficiency of the fuzzy controller obtained with the proposed method. The expression of the mean squared error (MSE) is the following:

$$MSE = \frac{1}{n} \sum\nolimits_{i=1}^{n} (X_i - Y_i)^2 \tag{4.11}$$

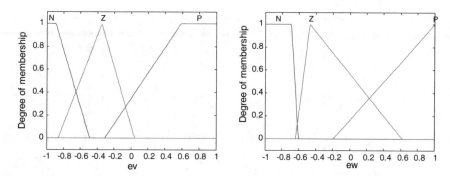

Fig. 4.14 Inputs of the fuzzy system of the autonomous robot problem

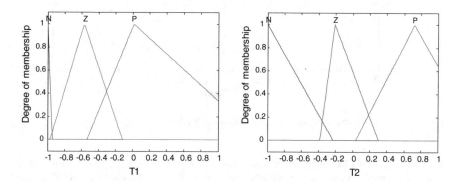

Fig. 4.15 Outputs of the fuzzy system of the autonomous robot problem

where MSE is the sum of squared errors, that is, the difference between the estimator and what is being estimated, X_i represents the reference value and Y_i represented the value produced by the system.

Finally, in Fig. 4.16 we can note in a graphical way the best result obtained by the proposed method, as we mentioned above is the best structure in 10 independents experiments and in comparison with other works, this proposed method uses a lower number evaluation functions than the other methods, in order to get a lower computational cost.

In order to complement Fig. 4.16, the best result obtained of the MSE is 0.0125, and with this result we can conclude that the proposed method has a good performance in the fuzzy control problem [28]. In addition, the STA has fewer evaluation functions in order to solve the autonomous robot mobile problem. In this case, the STA shows that the trajectory of the mobile robot (that is represented by the blue line follows) approximates well the desired trajectory that the problem has and is represented by the green line.

In Fig. 4.16 we can note that both lines (blue and green) are similar and the robot follows the trajectory with success according to the plot and this is very important

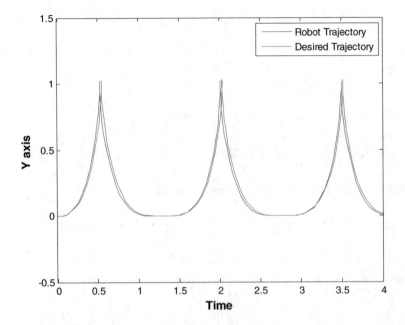

Fig. 4.16 Best experiment obtained by the String Theory Algorithm

because we can say that the proposed method also works with fuzzy control problems, where we need improve the structure of the Fuzzy Inference System by optimization.

References

1. J. Digalakis, K. Margaritis, On benchmarking functions for genetic algorithms. Int. J. Comput. Math. **77**, 481–506 (2001)
2. M. Molga, C. Smutnicki, *Test Functions for Optimization Needs*, unpublished
3. X.-S. Yang, *Test Problems in Optimization* (2010), arXiv, preprint arXiv:1008.0549
4. J. J. Liang, B.Y. Qu, P. N. Suganthan, Q. Chen, *Problem Definitions and Evaluation Criteria for the CEC 2015 Competition on Learning-bases Real-Parameter Single Objective Optimization* (2014)
5. W. Guohua, R. Mallipeddi, P.N. Suganthan, *Problem Definitions and Evaluation Criteria for the CEC 2017 Competition on Constrained Real-Parameter Optimization* (2017)
6. L. Zadeh, Fuzzy Logic = computing with words. IEEE Trans. Fuzzy Syst. **4**(2), 103–111 (1999)
7. M.A. Hossain, I. Ferdous, Autonomous robot path planning in dynamic environment using a new optimization technique inspired by bacterial foraging technique. Rob. Autom. Syst. **64**, 137–141 (2015)
8. M.L. Lagunes, O. Castillo, F. Valdez, J. Soria, P. Melin, *Parameter Optimization for Membership Functions of Type-2 Fuzzy Controllers for Autonomous Mobile Robots Using the Firefly Algorithm* (Springer, Cham, 2018), pp. 569–579
9. T. Lee, K. Song, C. Lee, C. Teng, Tracking control of unicycle-modeled mobile robots using a saturation feedback controller. IEEE Trans. Control. Syst. Technol. **9**, 305–318 (2011)

10. R. Martinez, L. Aguilar, Optimization of interval type-2 fuzzy logic controllers for a perturbed autonomous wheeled mobile robot using genetic algorithms. Inf. Sci. (Ny) **179**, 2158–2174 (2009)
11. C. Caraveo, F. Valdez, O. Castillo, A new metaheuristic of optimization with dynamic adaptation of parameters using type-2 fuzzy logic for trajectory control of a mobile robot. Algorithms **10**, 85 (2017)
12. Y. Kanayama, Y. Kimura, F. Miyazaki, T. Noguchi, *A Stable Tracking Control Method for an Autonomous Mobile Robot* (The NPS Institutional Archive, Calhoun, 1990)
13. J. Barraza, P. Melin, F. Valdez, C. González, *Fuzzy FWA with Dynamic Adaptation of Parameters*. (CEC, 2016), pp. 4053–4060
14. F. Olivas, G. Amador, J. Perez, C. Caraveo, F. Valdez, O. Castillo, Comparative study of type-2 fuzzy particle swarm, bee colony and bat algorithms in optimization of fuzzy controllers. Algorithms **10**, 101 (2017)
15. L. Rodriguez, O. Castillo, J. Soria, P. Melin, F. Valdez, C. Gonzalez, G. Martinez, J. Soto, A fuzzy hierarchical operator in the grey wolf optimizer algorithm. Appl. Soft. Comput. **57**, 315–328 (2017)
16. R. Larson, B. Farber, *Elementary Statistics Picturing the World* (Pearson Education Inc., 2003), pp. 428–433.
17. L. Rodríguez, O. Castillo, J. Soria, *Grey Wolf Optimizer with Dynamic Adaptation of Parameters Using Fuzzy Logic* (CEC, 2016), pp. 3116–3123
18. M. Lagunes, O. Castillo, J. Soria, Optimization of membership functions parameters for fuzzy controller of an autonomous mobile robot using the firefly algorithm, in *Fuzzy Logic Augmentation of Neural and Optimization Algorithms* (2018), pp. 199–206
19. G. Arslan, On a characterization of the uniform distribution by generalized order statistics. J. Comput. Appl. Math. **235**, 4532–4536 (2011)
20. E. Nabil, A Modified Flower Pollination Algorithm for Global Optimization. Expert Syst. Appl. **57**, 192–203 (2016)
21. M. Jamil, H.J. Zepernick, Levy flights and global optimization, in *Swarm Intelligence and Bio-inspired Computation* (2013), pp. 49–72.
22. Lingaraj and Haldurai, A study on genetic algorithms and its applications. Int. J. Comput. Sci. Eng. **4**, 139–143 (2016)
23. X.-S. Yang, *Flower Pollination Algorithm for Global Optimization* (2012), arXiv:1312.5673v1
24. O. Castillo, H. Neyoy, J. Soria, P. Melin, F. Valdez, A new approach for dynamic fuzzy logic parameter tuning in Ant Colony Optimization and its application in fuzzy control of a mobile robot. Appl. Soft. Comput. **28**, 150–159 (2015)
25. B. Salah, Skin cancer recognition by using a neuro-fuzzy system. Cancer Inf. **10**, 1–11 (2011) (PMC.Web. 6 Dec. 2016)
26. P. Ochoa, O. Castillo, J. Soria, Differential evolution with dynamic adaptation of parameters for the optimization of fuzzy controllers, in *Recent Advances on Hybrid Approaches for Designing Intelligent Systems* (Springer, Cham, 2014), pp. 275–288
27. C. Peraza, F. Valdez, J. Castro, C. Oscar, Fuzzy Dynamic parameter adaptation in the armory search algorithm for the optimization of the ball and beam controller. Adv. Oper. Res. 1–16 (2018)
28. P. Melin, F. Olivas, O. Castillo, F. Valdez, J. Soria, M. Valdez, Optimal design of fuzzy classification systems using PSO with dynamic parameter adaptation through fuzzy logic. Exp. Syst. Appl. **40**, 3196–3206 (2013)

Chapter 5
Conclusions

Finally, we can mention two different types of conclusion, the first one is based on the proposed basic randomness method and the second one is the conclusion about the String Theory Algorithm.

In this book, we have analyzed a new approach to generate randomness for meta-heuristics, which is one of the main features used in the metaheuristics techniques and that used implicitly a stochastic model. This new way is based on the basic behavior of the String Theory that we can find in nature and physics laws according to its theory.

In addition in this book, we used three sets of problems in order to test the new way to generate randomness number based on the movement of Strings, the first set was the 13 traditional benchmark functions classified as unimodal and multimodal, the second set were more complex problems that were created for the CEC 2015 competition and included 15 benchmark functions and the third set of problem were the constrained problems of the CEC 17 competition. These problems were used for testing the three original methods in comparison with the proposed method, used in three different metaheuristics and finally the results were compared through a hypothesis test. Also we presented a sensitivity analysis in order to prove that the proposed method have a good performance even when the parameters are adjusted for the first set of benchmark functions, with this result we can conclude that the proposed method could be used in other problems when the parameters of the algorithms need to be optimized.

In addition and according with No Free Lunch theorems for optimization proved that there is no meta-heuristic appropriately suited for solving all optimization problems. For example, a particular meta-heuristic can give very promising results for a set of problems, but the same algorithm can show poor performance in an another set of different problems as we could notice in the simulation results.

© The Author(s), under exclusive license to Springer Nature Switzerland AG 2022
O. Castillo and L. Rodriguez, *A New Meta-heuristic Optimization Algorithm
Based on the String Theory Paradigm from Physics*,
SpringerBriefs in Computational Intelligence,
https://doi.org/10.1007/978-3-030-82288-0_5

It is important remember that we recommended use the proposed method when the problem has as minimum 10 variables or dimensions, in order to respect the nature and definition of the sine wave, in other words, we should have sufficient data to draw a perfect sine wave.

Finally, we can say that the new proposed randomness method based on the String Theory has demonstrated that it has a good performance, especially when the problems are complex as shown by the hypothesis test, this can be justified in some way by the behavior that some elements in the nature have and cause a sine wave behavior phenomena in real world systems.

The second phase of conclusions is totally about of the complete algorithm that we proposed (String Theory). In this book we presented a new algorithm based on String Theory, and we presented the main inspiration and equations that we used in order to represent the String Theory concepts as a metaheuristic and its main features according to the literature review.

In order to test the proposed method, we used a set of traditional benchmark functions. In addition, the results were statistically compared with three other metaheuristics (FPA, FA and GWO) where the newly proposed method proves that it has better performance than FPA and FA respectively. Finally, in comparison with GWO, we can conclude that String Theory Algorithm has better performance when the problem is more complex.

In addition, we presented a set of more complex benchmark functions, which are the functions presented in the CEC 15 competition, respectively. Also, we compared the results of the String Theory Algorithm with the FA and GWO algorithms and we can conclude that in the comparison with the FA in 30 and 50 dimensions respectively, both algorithms show similar performance and that both have better performance than the other in half of the total of benchmark functions that were analyzed. Finally, in the comparison with the GWO, the String Theory Algorithm, in general, has better performance in this set of benchmark functions (CEC 15).

Also in this book, we also presented an optimization problem that uses fuzzy logic in order to improve the controller of a robot. In this case, STA optimized the fuzzy controller of a robot autonomous mobile. String Theory Algorithm shows that with few evaluations is possible to obtain good performance and approximate the goal with a small mean squared error.

Finally, we can say that the new proposed String Theory Algorithm has demonstrated that it has a good performance with the set of three different study cases. These results were shown by the help of hypothesis testing. The good performance can be justified in some way with the fact that String Theory in theory can explain everything in the universe as being produced by the vibrations of these Strings and these techniques can be applied in computer science in order to improve the optimization problems.

Future Works

In this section we are presenting some future works that we can conclude about this book.

In the randomness proposed method, apply the basic randomness method in other metaheuristics as Particle Swarm Optimization (PSO), Bat Algorithm (BAT), Gravitational Search Algorithm (GSA) and other metaheuristics algorithms for optimization in order to test the performance of the proposed method.

Also it is important to mention that the proposed method could be compared with other techniques that generate random numbers in the metaheuristics, in this book we presented the comparison with random numbers, normal distribution and Levy flights.

In the String Theory Algorithm is important to mention that the number of parameters could be modified in order to improve the performance.

Perform a study of which values in the parameters are better for solve optimization problems.

Apply the dynamic adaptation of the parameter in the algorithm using fuzzy logic.

Apply the proposed method to problems of control systems and other benchmark plants about fuzzy inference systems.

Apply the proposed method to problems of optimization of artificial neural networks.

Index

© The Author(s), under exclusive license to Springer Nature Switzerland AG 2022
O. Castillo and L. Rodriguez, *A New Meta-heuristic Optimization Algorithm Based on the String Theory Paradigm from Physics*,
SpringerBriefs in Computational Intelligence,
https://doi.org/10.1007/978-3-030-82288-0

Printed in the United States
by Baker & Taylor Publisher Services